NUCLEAR INSIGHTS

VOLUME 3

NUCLEAR INSIGHTS:

THE COLD WAR LEGACY

VOLUME 3:
NUCLEAR REDUCTIONS
(A Technically Informed Perspective)

Alexander DeVolpi

2009

Dedication:
— to the officeholders and administrators of Barack Obama's presidential governance — from one of your earnest supporters: May you wisely and open-mindedly place high value on substantive experience, relevant knowledge, scientific methodology, technological feasibility, and environmental credibility.

[Historic Leadership Announcement About a Nuclear-Free World].
We committed our two countries to achieving a nuclear-free world, while recognizing that this long-term goal will require a new emphasis on arms-control and conflict-resolution measures, and their full implementation by all concerned nations.

–Joint statement by President Dmitry Medvedev of the Russian Federation and President Barack Obama of the United States of America (1 April 2009)

Version V3.A1c (October 2009)

EAN 978-0-9777734-4-2
ISBN 0-9777734-4-2

Front Cover Design: Laura M. Smith

Typesetting: WordPerfect®, PDF, and Adobe PhotoShop

Printer: Create Space/Amazon.Com

Cover Illustration: As the Barack H. Obama administration gradually takes office, memories and fears of the Cold War continue to fade. The first Volume's dark and foreboding cover in this trilogy depicts a violent thermonuclear-explosive test, one of many conducted and polluting the Earth's atmosphere. For Volume 2, the cover symbolizes a peaceful transition from military to civilian nuclear applications.

Volume 3's front cover is intended to reflect the greening of nuclear, especially its worldwide use in producing energy. Electricity-generating reactors are the very medium by which nuclear weapons can be and are being irreversibly destroyed. The cover graphically depicts a peaceful residential neighborhood, supplied by electricity from nuclear-power plants, which simultaneously produce clean and nearly endless energy. The triangular radura affixed to a nuclear-plant dome is an icon symbolizing its benign applications.

PRIMARY CONTENTS

VOLUME 3

TABLE OF CONTENTS
FOR PRELIMINARY PAGES

PREFACE FOR VOLUME 3

This third volume, encompassing Chapter VII of *Nuclear Insights: The Cold War Legacy*, is devoted to nuclear reductions. As in the two preceding volumes of *Nuclear Insights,* citations and many extraneous details have been omitted. They can be found in corresponding chapters of the co-authored *Nuclear Shadowboxing: Contemporary Threats from Cold War Weaponry.*

Whereas Volumes 1 and 2 of *Nuclear Insights* have supplied a history of nuclear weapons development and assessed the post-Cold-War legacy, this third and concluding Volume is an informed longer-term perspective: It explains how strategic stability can be maintained and improved by systematically reducing nuclear arsenals. It also constitutes a (substantive, rather than political) history of the modern movement towards their elimination.

While each volume has its own index, this one appends a consolidated table of contents, which amounts to a topical overview — a roadmap — of all three volumes. This is especially useful for tracking recurrent themes in nuclear-arms control that have a history, a current role, and a future. Ballistic-missile defense is an appropriate example that transcends all three Volumes.

Reader comments, corrections, advice, and criticism are welcome. With the advent of CreateSpace/Amazon.com "books-on-demand," it is much easier to make even minor adjustments for future editions without wasteful inventory.

****

Relevant expertise is important in analyzing the complex and multifaceted nuclear panorama: It involves not only technological factors, but also political, diplomatic, strategic, military, economic, and psychological dimensions. That's why a detailed summary of authorship credentials is provided at the end of this Volume. This book has evolved as a multi-decadal product of experienced technologists who are cognizant of applicable parameters and circumstances. Combined, we have had as much relevant first-hand experience as any who advise the implementation and verification of nuclear-weapons reductions.

More specifically, we (a diminishing breed, even among nuclear physicists) are among the very few who have gazed upon, handled, or exploded nuclear weapons of one or both of the nuclear superpowers. One or another of us has seen just about every nuclear weapon in our arsenals, as well as their support and delivery systems. We have visited in a technical capacity nearly every nuclear-weapon related facility in the United States and the former Soviet Union, and many weapon-related facilities in Great Britain, France, and a few in China.

One or more of us have been to almost every national or international facility or military installation involved in arms control and treaty verification — Europe, North America, and throughout Asia. At some time or another, we have had personal opportunities to experience first-hand most nuclear and non-nuclear laboratories, facilities, and institutions devoted to nuclear engineering, development, and production throughout the world.

Moreover, we have directly engaged in analysis, experiments, and field work for inspection and verification of armaments. We have worked with, advised, or been involved with numerous activist NGOs. We carried out military duties in our respective armed services.

Most importantly, we had enough training, experience, and hands-on practice in the basic science, technology, and applications to understand the underlying functioning and flaws of human-derived mechanisms.

Also relevant is our long-term experience with quality control and product standards; these are formative attributes widely and conscientiously practiced in nuclear experiments and nuclear engineering — witness the long-term reliability of both reactors and weapons. Coupled with this is an earned appreciation for statistical properties, random and systematic, that govern balanced assessments of fabricated equipment and processes. Appreciation of limitations imposed by nature's constraints cannot be fully ingrained in the classroom; it is something largely attained by intensive hands-on familiarization. Part and parcel of this process is humble awareness of human flaws and error.

Although you can find many substantive, excellent, and scholarly renditions of the nuclear heritage, none — I assure you — have been prepared with the same depth of intimate and relevant experience. These are key reasons why this Volume can be billed as an informed perspective about navigating through potential pitfalls to reach a nuclear-stable world.

Chapter VII:

NUCLEAR REDUCTIONS/ DISARMAMENT

(The Future?)

*W*ith the implacable hostility between superpowers gone and the Soviet totalitarian empire dissolved, most obstacles to nuclear arms verification have evaporated — thus releasing the paramount drag on arms reductions and disarmament. Both the United States and Russia now share common interests: avoiding nuclear conflict, reducing military spending, and preventing leakage of nuclear material and knowledge. Preserving these shared interests of the 21st century would seem to be a high priority.

This retrospective about the nuclear-arms race has policymaking value if applied to an admittedly difficult task for the new millennium: time-bound movement toward nuclear disarmament. Progress in this direction could help mitigate public apprehension about nuclear conflicts, military spending, proliferation, and terrorism.

Yet, from an international view, lack of significant superpower progress in disarmament warrants criticism, aptly summarized not so long ago by a Canadian representative to the UN:

> [As] they did during the Cold War, [the two principal nuclear powers continue] to ignore the language of the Nuclear Non-Proliferation Treaty (NPT), which says that the nuclear powers must "pursue negotiations in good faith for nuclear disarmament." They continue to hold in scarcely disguised contempt the unanimous 1996 World Court advisory opinion that said the nuclear weapon parties to the NPT had incurred a legal obligation to actually achieve nuclear disarmament. They continue to insist on the "right" of first use of nuclear weapons. And they continue to keep thousands of nuclear warheads on hair-trigger alert.

Indeed, strategies and threats of mutual assured destruction remain more than a decade later. Risk of self-annihilation has become second nature to nuclear-armed nations: Despite comforting assurances by government officials, only a few political dangers have diminished, while many hazards of nuclear weaponry persist. In this Volume, it is shown that residuals of the Cold War can be reduced in such a way as to bolster national security while providing public relief from exorbitant risk and taxation. This final Chapter is unavoidably filled with policy and technical particulars, but readers can skip overly detailed or complex topics and still understand the gist of our analysis.

With the new millennium well underway, the five major powers have made little progress toward reducing their dependence on nuclear weapons, as one can surmise from the discussion below.

*What you will **not** find below is a discourse about "abolishing" nuclear weapons. It's an apt goal for political figures, a few of whom share responsibility for the current state of affairs, but too impractical for those of us who have in-depth, hands-on experience with nuclear weaponry and arms-control verification. Abolition is an admirable goal, but it should not jeopardize more manageable reductive measures that could greatly detoxify the prevailing existential threat.*

A. NUCLEAR-POLICY DEBATE

Arms-control policy analysts recognize that "the most conspicuous characteristic of the Cold War's military confrontation" has been "halted and reversed," yet the "physical threat that these [nuclear] weapons pose" has not been sufficiently altered. "The basic structure of plans for using nuclear weapons appears largely unchanged...." with "both sides apparently continuing to emphasize early and large counterforce strikes and both remaining capable" — despite reductions in numbers and alert levels — of rapidly bringing nuclear forces "to full readiness for use." Moreover, nuclear nations evidently plan to maintain their weapons for the indefinite future.

The Soviet collapse brought many to think the nuclear threat had vanished and nuclear deterrence would take care of any residual threats. Nobel Peace laureate Joseph Rotblat, though, pointed out that "neither of these beliefs is justified" as the 21st century gets underway. Two new nuclear-weapon states were at least once "poised for a military showdown over Kashmir," and the Clinton administration had "made explicit mention of the use of nuclear weapons in response to an attack with chemical or biological weapons." Known as "extended deterrence," this *de facto* strategy, expanded by the G.W. Bush administration, "implies the indefinite existence of nuclear arsenals."

Originally U.S. policymakers thought that the atomic bomb would force Soviet leaders to bend to American desires. But they quickly found the political utility of nuclear threats to be rather limited — because of potential Soviet retaliation in kind, and because of the enormous risk-taking aspects of confrontation. American nuclear arsenal expansion spurred a massive Soviet response. Although nuclear aggrandizement was kept in check (nuclear weapons have been a great equalizer), the United States became vulnerable, for the very first time, to devastating attacks from a foreign power. Moreover, small — otherwise weak — countries became capable of endangering much larger and more powerful nations; even now, the possibility that small nations could develop deliverable nuclear weapons has created severe consternation in the United States.

While admittedly not possible to disinvent* nuclear weapons, authors of a well-formulated book, *The Nuclear Turning Point,* asserted that "the taboo on their use can be strengthened." A moderately successful multinational norm has developed against the use of, or explicit threat to use, nuclear weapons. A weak international norm has minimized proliferation. As of 2009, only four nations had not signed the NPT, while nuclear-weapon-free zones have spread to Latin America, Africa, the South Pacific, and Southeast Asia. "Negative security assurances" (promising not to initiate nuclear warfare) have been offered to parties of the NPT, and "positive security assurances" (pledges of military or diplomatic assistance) have been made to nations that might be threatened with aggression involving nuclear weapons.

A Post-Cold-War Warning. The Committee on International Security and Arms Control (CISAC) of the National Academy of Sciences, in their 1997 report "The Future of U.S. Nuclear Weapons Policy," deemed "unacceptably high" the dangers of initiating nuclear war by error (based on false warning of attack) or by accident (a technology failure). Dangers remain as a result of continued reliance on nuclear retaliation. Because "highly survivable basing modes" make it "impractical to execute a disarming first strike," CISAC recommended that military strategy should be confined to a "core function" of deterring *nuclear* attack or resisting nuclear coercion.

The CISAC report advised a two-part modification of nuclear-weapons policies: (1) "near- and mid-term force reductions" — together with accompanying changes in nuclear operations. declaratory policies, and measures to increase the security of nuclear weapons and fissile materials worldwide, and (2) "a long-term effort to foster international conditions in which the possession of nuclear weapons would not longer be seen as necessary or legitimate for the preservation of national and global security." Advocated also by CISAC was "a regime of progressive constraints" that would expand arms control, retain existing treaties, restrict nuclear

---

* To rescind the invention or existence of: "The atomic bomb cannot be disinvented."

operations, enhance weapons and materials safeguards, strengthen confidence-building measures, eliminate continuous-alert practices, revise operational doctrine, expand security assurances, strengthen nonproliferation, universalize nuclear-arms reductions, improve verification, and expand transparency measures.

The preceding agenda was based on recognition that military security of cooperating nations would benefit from verifiable and permanent reduction of nuclear weapons.

The Generals Speak Out. General George Lee Butler, the last SAC commander and once a strong-nuclear-posture protagonist, has since declared that "The Cold War lives on in the minds of those who cannot let go the fears, the beliefs, and the enmities born of the nuclear age [including] the persistent belief that retaliation with nuclear weapons is a legitimate and appropriate response to post-Cold War threats posed by weapons of mass destruction...."

With that traumatic period having become "history," author Jonathan Schell, who interviewed flag officers, came to the conclusion that "The chief political question is whether nuclear proliferation can be stopped and reversed [at a time] when current nuclear powers [continue to] declare by their actions as well as their words that ... nuclear bombs are indispensable instruments of power." As he puts it, "Can we still remember that to destroy hundreds of millions of human beings is an atrocity beyond all history?"

Schell also wrote:

The fact is that the public at large, enjoying a reprieve from immediate, universal terror bestowed by the end of the Cold War, is not paying much attention to the nuclear question.... The combination of comparatively low nuclear danger and high opportunity to solve the nuclear dilemma is new.

It may be that the initiative for such a radical challenge to the status quo must be taken by the public, as has happened so often in the past, and especially in the United States.... The leadership might come from some unexpected quarter.... If [the new generation] members do not feel the urgency to escape the nuclear danger that some of its parents felt, neither has it developed the deep attachment to nuclear arms often found among the parents, including most of the governing class.

One factor retarding nuclear disarmament is a common misperception about national greatness: Schell quotes a Russian general: "A great power ... could be identified by six characteristics: a large territory, a large population, powerful conventional forces, a strong economy, membership in the UN Security Council, and possession of a strategic nuclear arsenal."

General Butler was an insider who became an arms-control booster; he started to promote the abolishment of nuclear weapons. In stating reasons for favoring abolition, Butler placed particular emphasis on the cessation of superpower hostility: "People who say to me that the elimination of nuclear weapons is utopian have somehow managed to completely ignore the fact that the end of the Cold War was a far more utopian prospect than eliminating nuclear weapons now." Moreover, he noted that "95 percent of the nations of the world are already nuclear free."

General Butler personally participated in what former Arms Control and Disarmament Agency (ACDA) director Kenneth Adelman labeled the "holy war," referring to surrogate arms-control negotiations "for the nuclear battle that never transpired." That experience left Butler with a less sanguine perspective, "Deterrence was really a uniquely Western concept."

Is the glass half-full or half-empty? Was the absence of nuclear battle a consequence of deterrence — or of negotiations based on realism?

To those who insist that deterrence prevented World War III, General Butler thought they were "seeing that proposition through a Western lens." When asked for his explanation on why there has been no third world war, part of his answer was

the trauma of the Cuban missile crisis was so great that both sides came to accept certain unwritten rules of the game [which] had already been taking shape. They boiled down ultimately to respect for spheres of hegemony, or influence.... There was no talk of deterrence in those critical thirteen days. What you had was two small groups of men in two small rooms, groping frantically in the intellectual fog, in the dark, to deal with a crisis that had spun out of control.

Another argument put forth by Butler: "For the statement that nuclear weapons prevented World War III to be true, it would first have to be the case that the Soviet Union had a compelling urge to launch an aggressive war against the West. At this moment [1998], we have yet to find evidence for that."

Nuclear weapons are irrational devices. They were rationalized and accepted as a desperate measure in the face of circumstances that were unimaginable. Now as the world evolves rapidly, I think that the vast majority of people on the face of this earth will endorse the proposition that such weapons have no place among us. There is no security to be found in nuclear weapons. It's a fool's game.

General Butler summed up his experience with the doctrine of deterrence:

the more [policymakers] tried to make the nuclear threat operational, the more they undermined the very stability they were trying to produce.... while it was true that nuclear weapons could not be disinvented, safety from nuclear danger would steadily increase as the world proceeded along the path from fully alert to fully disassembled arsenals.

In order to contrast conflicting viewpoints, Schell categorized proponents and opponents as "abolitionists" and "possessionists." General Charles Horner, who too became an abolitionist, declared, "The nuclear weapon is obsolete; I want to get rid of them all." This "deeply considered opinion" partly resulted from his experience in commanding allied air forces in the 1990-1991 Gulf War: "Nuclear weapons are such a *gross* instrument of power that they really have no utility."

Robert McNamara's well-known change of heart about nuclear retaliation was summarized by Schell: "To use nuclear weapons against a nuclear-equipped opponent of any size at all is suicide, and to use them against a nonnuclear-equipped opponent is ... immoral."

Along with former Senator Alan Cranston and former CIA Director Stansfield Turner, McNamara articulated the view that as long as nuclear weapons exist, "sooner or later, it's only a matter of when, not if, they will be fired again — by a terrorist, by accident, or by design. The strategy of deterrence ... cannot be foolproof forever." These former government officials became convinced the peril was graver because of chaos and shortage of resources in Russia. This, they said, has led to "a lack of proper command and control over nuclear weapons," leaving them "open to unauthorized use, an accidental use," or "terrorists ... might buy or bribe or steal their way to getting nuclear weapon components and capacities...." Senator Cranston cited polls showing 80 percent of Americans had become to believe that the world would be better off without nuclear weapons.

General Colin Powell, who later became President Bush's Secretary of State, is quoted as saying that after he saw a plan for nuclear-strike options, he no longer had any doubts "about the [impracticality] of nukes on the field of battle."

WMD Fright. Nuclear, biological, and chemical (NBC) are the dreaded threesome (the latter two are better labeled "weapons of indiscriminate casualty"). Some analysts fear that potential WMD dangers were not being taken seriously. For example, a prominent hard-liner, Colin Gray, considered them "a major menace." He was admittedly "reared on the thoughts" of Brodie, Wohlstetter, Kissinger, Schelling, and Kahn, all defense analysts who emphasized worst-case scenarios.

Despite by most nations banning and destroying chemical and biological weapons, Gray considered international control to be "impractical." Lacking the "nuclear taboo" and effective external control, Gray predicted that BC agents will cause a "shock" in the 21st century. Like other hard-liners who flourish on anxieties and enemies, Gray was distressed that "biological and chemical weapons pose some unique [strategic and analytical] challenges." His nemesis consists of "polities and substate groups" that will "figure out strategically effective ways to use BC agents for coercion." To deal with these, Gray proposed "counterforce" as essential to NBC antiproliferation, including threats, special operations, conventional weapon strikes, and "even precise nuclear assault, backstopped by several layers of air and ballistic missile defenses."

Gray also wrote that the ABM treaty "has no strategic relevance now" (because there was no strategic imbalance and because it prevented "really effective missile defenses"). He judged that there is too much confidence in deterrence, which he considered "wonderful" when (or if) it worked. To Gray, the existence of nuclear weapons is a *condition*, not a *problem*: They represent a "vital means" for "imposition of damage in battle" and for achieving a direct "strategic effect" on "policymaking processes." Though favoring nuclear weapons because they still "serve a variety of political goals," he acknowledged (perhaps warned) that nuclear wars can happen — from carelessness, accident, malfunction, miscalculation — plus "(small) nuclear wars waged by belligerents who fully intended to wage them." To Gray, "we are entirely in the realm of guesswork when we pass judgment on the prospects for continuation of the long nuclear peace."

To those who have tracked G.W. Bush administration policies, Gray's viewpoints in the last two paragraphs might resound with familiarity.

In order to realistically assess inherited WMD risks, dangers of military conflicts must be differentiated from risks and consequences of terrorist attacks. While chemical weapons have been used in inter-nation and intra-nation warfare, their known role in sub-national (terrorist) activities has been extremely limited, the primary pre-millennium exception being a terrorist group in Japan that experimented with biological agents.

The 9/11 suicidal crashes into the World Trade Center and Pentagon demonstrate that there has indeed been an erosion of what was once an intrinsic barrier (self-preservation), which previously deterred WMD acquisition and use by sub-national groups.

The G.W. Bush administration pledge to "combat" WMD emphasizes counter-proliferation and counterterrorism by aggressive interdiction. The President relegated nonproliferative and mitigative measures to be secondary and tertiary "pillars" of national strategy.

The "war against terrorism" became the central focus of Bush-administration policy. Despite militant and encompassing nomenclature, the counterterrorism campaign has not had the same magnitude or impact as did Cold War preparedness; nor do efforts to forestall terrorism yet employ the same means or goals as in the darker days of superpower conflicts. Although outsider acquisition and/or use of weapons of indiscriminate casualty would represent a defeat for national and international countermeasures, terrorist attacks cannot be fully repressed by classic deterrence or campaigns. Their impact can be certainly minimized, by the elimination of existing WMD stockpiles, as well as by bolstering what are otherwise poorly regulated conventional-arms exports.

Different Persuasions of Nuclear Deterrence

In the post-Cold-War era, attached to the classic role of nuclear deterrence have been new meanings, especially with the advent of civilian "strategists" outside government bureaucracies. Variations include unstable deterrence, passive deterrence, existential deterrence, extended deterrence, and virtual arsenals.

Unstable Deterrence. Apprehension has been increasing among those familiar with or have been invited to observe Russia's inherited nuclear forces. Its 1999 command and control system, assessed in of *The Nuclear Turning Point*, had been "deteriorating in ways that could jeopardize the ability of the country's central authority to control nuclear weapons." This was "deeply worrisome" because "U.S. and Russian command and control systems could [inadvertently] interact in dangerous and unstable ways...."

This problem was compounded because, going into the millennium, both nuclear superpowers "maintain and regularly exercise a capability to launch on warning thousands of nuclear warheads after a missile attack is detected." The arsenals remain ready for nearly instant use. War plans have changed little in structure, scope or detail. Force structures and doctrines originally designed to deter deliberate large-scale nuclear attacks and conventional war in Europe are obsolete and jeopardize security.

The greatest contemporary danger to the United States, concluded a veteran NGO analyst, became "a Russian attack resulting from error." Although Russia officially denies serious problems of nuclear control, there were reported to be "abundant signs of technical and organizational decay and human duress. Designers of command systems in Russia went to extraordinary lengths to ensure strict control over nuclear weapons, a core value of Soviet political and military culture."

Yet, according to U.S. intelligence reports for Russia, there have been

frequent malfunctions of command-and-control equipment and intermittent spontaneous switching to a combat mode for no apparent reason. Power to key nuclear weapons installations have been cut off numerous times for nonpayment of bills, and on seven occasions during autumn 1996, operations at some nuclear facilities were disrupted because thieves were "mining" communications cables for valuable metals.

Also suspected to be at risk have been special launch-enable codes, which "may be distributed fairly widely to alternative command centers." Moreover "some [Russian] submarine crews may possess autonomous launch capability for the ballistic missiles on board." Weak links such as these imply that one or a few individuals could conceivably seize control of one or more strategic weapons and launch them without governmental authorization. Presumably these weaknesses have been since remedied.

Nevertheless, dangers of mistaken, unauthorized, or accidental attack nowadays exceed the risks of a deliberate government decision to attack (although, frankly, all of these risks are quite small). Moreover, retaliation has little operative value in deterring these dangers; in fact, precipitous response to an unprovoked incident could propel the situation out of control. The quandary is that a very small risk could have intolerably severe consequences.

Nuclear Turning Point authors were particularly concerned about the Russian early warning network, which was constructed by the former Soviet Union to detect a ballistic missile attack; it is "perhaps the most neglected component of the strategic posture."

[Russia] lacks "dual phenomenology" (detection using two different types of sensors) for many potential attack corridors and thus has no way to check rapidly a potentially false report issued by a single sensor. [Control] over nuclear weapons could splinter along political fault lines.... The General Staff's direct access to the launch codes enables its members to initiate a missile attack with or without the permission of political authorities.

   Within just a few minutes after receiving the order to fire, a large fraction of the U.S. and Russian land-based rockets armed with 2000 and 3500 warheads, respectively, would begin their 25-minute flight over the North Pole to their wartime targets. Within 20 minutes after ordering the attack, strategic submarines would empty their launch tubes. In sum, daily alert postures enable the nuclear superpowers to launch within 20 minutes about 5000 nuclear weapons aimed at military installations and cities. Given the weakening control over Russian forces, the unintentional use of nuclear weapons is a more serious threat than is a failure of deterrence leading to a calculated act of nuclear aggression. The United States and Russia are running a growing risk of stumbling into an inadvertent war in order to deter a vanishing risk of deliberate attack.

   To meet this demand for comprehensive target coverage, both the United States and Russia plan to launch on warning — launching a massive retaliatory salvo after detecting an enemy missile attack but before the incoming warheads arrive 15 to 30 minutes later.

That (outdated) hypothetical scenario describes how an irreversible nuclear exchange could have unfolded without being under a real attack.

Take a common analogy from highway driving: Just think how easy it is to reflexively apply your brakes if spooked while driving fast at night. You might later wish you weren't so hasty in your reaction, but usually the consequences will be limited. Not so if a ballistic-missile exchange commences: only 15 to 30 minutes is available to respond effectively; that allows just a few minutes to assess the warning information and a few minutes for top-level decisionmaking. A strong incentive therefore exists to fire missiles from comparatively vulnerable fixed ICBM silos and home-ported submarines before they could be pulverized by incoming warheads. Although the respective governments issue reassuring statements, the hypothetical nightmare scenario is difficult to shake off.

Here is another factor that some influential Americans evidently still do not understand: It is not possible for outsiders to dictate Russian perceptions; both nations view security through their own sociological prisms, which are blurred by indigenous historical impressions.

While one side might find it useful to approach national security with inflexibility, the other side might find a non-negotiable stance to be destabilizing or threatening. Presidential decision directives that influence military policy are usually condensed from strategic reviews. Once embodied in leadership directives, coexistence with nominally unstable deterrence became a hallmark of the Cold War.

Passive Deterrence. Ongoing policy debate counsels that a credible post-Cold-War role for nuclear weapons is passive deterrence — defined as an implied, less overt strategic posture derived from a much smaller nuclear arsenal.

Related to passive deterrence is "self-deterrence," a phrase used by Admiral Stansfield Turner, who served as CIA Director under President Jimmy Carter. Turner advised that nuclear attack was deterred by fear of reprisal in kind. He is quoted as saying "There is no foreign policy objective ... so threatened that we would employ nuclear weapons and accept the risk of receiving just one nuclear detonation in retaliation."

The Brookings *Nuclear Turning Point* recommends the "overriding goal" of U.S. nuclear policy should be "to prevent the use or threat of use of nuclear weapons against the United States and its allies and to prevent the further spread of the weapons." Passive deterrence would fit that goal if the magnitude, texture, and status of residual arsenals were configured to discourage reliance on nuclear weapons.

The Cold War standoff between East and West was representative of both active and passive deterrence, depending on the ongoing balance or imbalance of nuclear forces; another example has been Israel's reputed nuclear arsenal which discourages a massive invasion. The strategic impasse between India and Pakistan has been an illustration of passive deterrence in a more contemporary situation.

The G.W. Bush administration feared that "rogue" states who acquire their own WMD will not be deterred by nuclear weapons. Just the opposite case was presented in 2002 by a political scientist: States like Iraq, North Korea, and Iran are indeed "much weaker militarily" than the nuclear powers; yet their leaders can easily see that any aggression they might initiate has been or would be "repulsed quickly and effectively by stronger, better-equipped" forces.

Nor have nuclear weapons translated into actual usable "offensive power." For instance, India — despite its nuclear demonstration decades ago — has not been able to achieve its goals in Kashmir, and nuclear-weapon states did not prevail in

Vietnam. Even if Iraq had acquired nuclear weapons, it would not have been able to recapture Kuwait or take Iran. With or without nuclear weapons, North Korea is deterred by a number of geopolitical incentives and pressures from launching an attack on any nation.

If "rogue leaders" obtained or developed nuclear weapons, it is unrealistic to expect them to be passed on to terrorist groups or nations: "They would not want to part with even one" of the few they could acquire. "Moreover, [homemade] weapons would be car-size and cumbersome, not terrorist-convenient."

A strategy of pre-emption, tempting as it is, has not provided much in the way of modern strategic gains. The 1981 Israeli attack on an Iraqi nuclear reactor was a least partly counterproductive, the Bay of Pigs invasion failed miserably, and the "successful 1982 invasion of Lebanon turned into a disaster for Israel." If one ignores the unfathomable and unchallenged gambits of superpower confrontation, passive deterrence worked well enough and has continued to function effectively so far.

Existential Deterrence. In their 1997 report noted above, CISAC contended "As long as nuclear weapons exist, this very existence will exert a deterrent effect — *existential deterrence* — against unrestricted conventional war among the major powers...." Nuclear weapons, they agreed, cannot be "uninvented."

Existential deterrence prevails when the existence (or potential existence) of nuclear weapons adds caution to the conduct of international affairs. This has been described by Jonathan Schell as a form of "doctrinal limbo," consisting of "an aura of menace that inheres in nuclear arsenals and radiates from them, whether or not their possessors make specific, defined threats or adopt elaborate targeting plans."

Whereas "passive deterrence," discussed earlier, is largely associated with restraints that keep nuclear-weapons states from attacking each other or being attacked by conventional forces, "existential deterrence" applies to scenarios that involve potential aggression against any nuclear-armed state.

Needless to say, existential deterrence is of little help in countering non-state terrorism, which has presented new and diffuse threats: Terrorists seem to revel in their ability to tweak the tail of nuclear lions.

Extended Deterrence. Some policymakers have considered nuclear weapons to be deterrents extended to chemical or biological warfare. It is plausible that, during the Persian Gulf War of 1990-1991, Iraq was deterred from using such weapons, but the nuclear threat is unworkable in most small-scale conflicts; it could not become a universal principle. Extending the threat to deter conventional conflicts has also been found to be of limited practical policy value. Moreover, wholesale retaliation is ignored by stateless terrorists who do not normally present an identifiable target.

Except for "an extreme circumstance of self-defense," the threat or use of nuclear weapons is contrary to international law. Explicitly, the International Court of Justice held in 1996 that

> [The] threat or use of nuclear weapons would generally be contrary to the rules of international law applicable in armed conflict, and in particular the principles and rules of humanitarian law. However, in view of the current state of international law, and of the elements of fact at its disposal, the Court cannot conclude definitely whether the threat

or use of nuclear weapons would be lawful or unlawful in an extreme circumstance of self-defense, in which the very survival of a State would be at stake.

A UN General Assembly resolution in 1961 declared that the use of nuclear weapons is

contrary to the spirit, letter and aims of the United Nations and, as such, a direct violation of the Charter of the United Nations ... contrary to the rules of international law and to the laws of humanity... and ...a crime against mankind and civilization.

Despite practical and legal limitations, extended deterrence probably best describes the military-action plans of most nuclear-weapons states.

Strategic Deterrence. Much the same is the concept of strategic deterrence embodied in U.S. doctrine for nuclear operations, as of 2005. The definition is directly quoted below including emphasis selected by the Joint Chiefs.

Strategic deterrence is defined as the prevention of adversary aggression or coercion that threatens vital interests of the United States and/or our national survival. **Strategic deterrence convinces adversaries not to take grievous courses of action by means of decisive influence over their decision making.** Deterrence broadly represents the manifestation of a potential adversary's decision to forego actions that he would otherwise attempt. Diplomatically, the central focus of deterrence is for one nation to exert such influence over a potential adversary's decision-making process that the potential adversary makes a deliberate choice to refrain from a course of action. **The focus of U.S. deterrence efforts is therefore to influence potential adversaries to withhold actions intended to harm U.S.' national interests.** Such a decision is based on the adversary's perception of the benefits of various courses of action compared with an estimation of the likelihood and magnitude of the costs or consequences corresponding to these courses of action. It is these adversary perceptions and estimations that U.S. deterrent actions seek to influence. Potential adversary decision making in the face of U.S. deterrent actions is also influenced by their strategic culture, idiosyncrasies of decision mechanisms and the leader's decision style, and leadership risk tolerance.

Supplementing this is a more explicit statement drawn from the same document:

Declaratory Policy. The U.S. does not make positive statements defining the circumstances under which it would use nuclear weapons. Maintaining U.S. ambiguity about when it would use nuclear weapons helps create doubt in the minds of potential adversaries, deterring them from taking hostile action. This calculated ambiguity helps reinforce deterrence. If the U.S. clearly defined conditions under which it would use nuclear weapons, others might infer another set of circumstances in which the U.S. would not use nuclear weapons. This perception would increase the chances that hostile leaders might not be deterred from taking actions they perceive as falling below that threshold. [In the past, when North Atlantic Treaty Organization (NATO) faced large Warsaw Pact conventional forces, the U.S. repeatedly rejected calls for adoption of a "no-first-use" policy of nuclear weapons, since this policy could undermine deterrence. The U.S. countered such calls by stating that it would not be the first to use force....]

The bracketed explanatory comment, drawn from a different part of the document, demonstrates how words can be engineered so as to avoid explicit determination to use nuclear weapons when the circumstances call for them.

Elsewhere in the document, it made clear that nuclear operations were not limited to *security*: Nuclear weapons could also be used to "dissuading adversaries from undertaking programs or operations that could threaten U.S. *interests* or those of our allies and friends." More explicitly, "the decision to use nuclear weapons is driven by the *political objective* sought." (That emphasis was added in order to accentuate the candid acknowledgment that political interests, not national security, are the driving force in keeping a versatile and operable nuclear arsenal. It is not clear, however, if these are domestic or international political interests.)

Another citation from the U.S. doctrine merits attention at this point:

> [While] the belligerent that initiates nuclear warfare may find itself the target of world condemnation, **no customary or conventional international law prohibits nations from employing nuclear weapons in armed conflict....**
>
> **The law of armed conflict does not prohibit nuclear weapons use in armed conflict** although they are unique from conventional and even other WMD in the scope of their destructive potential and long-term effects.

To reach these conclusions, the Joint Chiefs — directed by the G.W. Bush executive branch — deliberately disregarded the International Court of Justice and the UN General Assembly.

Virtual Arsenals. Just short of completely eliminating nuclear weapons are some proposals for conversion to "virtual arsenals," that is, vastly reduced but not abolished stockpiles of nuclear weapons. By getting all nuclear-weapons states involved, systematic reductions would take place according to some agreed formula. Each weapon state would retain the knowledge, materials, and components to reassemble their nuclear arsenals in a relatively short period of time.

The concept has the merit of being able to grapple with the underlying fact that many nations know how to build nuclear weapons: such knowledge cannot be undone. If virtual arsenals were considered a transition stage toward outlawing nuclear weapons, it might achieve acceptability. Some individuals who favor outright nuclear abolition are uncomfortable with the concept of virtual arsenals.

In any event, nuclear rearmament is always a possible response to adversarial breakout from any gradual-reduction regime. The greater the degree of reduction, the longer it would take to reconstitute Armageddon-size arsenals.

The Cold War arms race featured an intense buildup and competition to gain or negate perceived nuclear advantage. Warheads were developed that had higher yields, better yield/weight ratios, lighter throw-weight, smaller size, better function, improved safety, enhanced radiation, etc. The bewildering array of weapons-delivery platforms included intercontinental missiles, aerial bombs, artillery shells, torpedoes, depth charges, demolition mines, and cruise missiles. These arsenals were designed for warfighting, not just deterrence. Thus, virtual arsenals (at much reduced levels) are a way of hedging so as to reduce the types of risk inherent in the Cold War arms race and its current legacy. To paraphrase: Old weapons need never die; they can just fade away.

Because of cost and dangers associated with large inventories of nuclear weapons, Russia and the United States would both benefit from mutual reductions. Current arms-control agreements and national means of intelligence collection already allow verification of reductions in strategic delivery systems, even in the absence of START II implementation or a START III treaty. Senator John Kerry in 1998 suggested that "the U.S. could safely reduce from the 6000 deployed strategic nuclear weapons down to the proposed START III levels of 2000-2500 ... because the U.S. has far more nuclear weapons than we need."

Kerry reasoned that,

This deployed arsenal no longer serves our national security interests, and it is provoking Russia to maintain an arsenal that undermines our national security interests... The several thousand nuclear warheads on Russian soil are the gravest, most imminent threats to the security of the United States.

Senator Kerry unsuccessfully attempted to amend the fiscal-2000 Defense Authorization bill so that additional nuclear-delivery systems could be retired before the Russian Duma ratified START II. He explained that the remaining warheads "are more than enough — many times over — to destroy any nation, anywhere, anytime, that threatens us. And the diversity of our triad — nuclear weapons on air, land and sea — protects us against the risk of a first strike destroying our capacity to retaliate."

The Council for a Livable World gave seven reasons to support the Kerry amendment: (1) Neither Russia nor the United States still needs huge stockpiles of nuclear weapons, a position upheld by the Pentagon; (2) substantial cost savings ($20 billion) will result from nuclear weapons cuts; (3) the savings can be applied to other Pentagon priorities; (4) reductions might jump-start the stalled nuclear reduction process; (5) the 1991 unilateral reductions carried out by Bush and Gorbachev provide a positive precedent; (6) U.S. security decisions should not simply be based on what happens in the Russian Duma; and (7) dismantling excess nuclear warheads will provide remaining nuclear weapons with a source of tritium, which is becoming scarce (discussed in later pages).

The United Kingdom, setting out a comprehensive nuclear-disarmament policy following a 1998 strategic review, noted that their participation in multilateral negotiations could include:

reduction in the size of our deterrent; greater transparency about our nuclear and fissile material stockpiles; placing fissile material no longer required for defence purposes under international safeguards; reprocessing of spent fuel from the defense Chapelcross reactors under international safeguards; beginning a national historical accounting for fissile material produced; and beginning a programme to develop UK expertise in verifying the reduction and elimination of nuclear weapons.

The concept of verifiably reducing nuclear weapons while retaining the knowledge and materials to rebuild them if needed (that is, virtual arsenals) is an approach consistent with the ongoing natural erosion of stockpiles. However, little attention was given by the G.W. Bush administration to reductions in strategic weapons. Well into 2009, the Obama presidency would be just getting into its nuclear-strategy reviews.

Along a different vein, another rationale has been invoked for virtual arsenals: A nation could have a nuclear capability even though it has not declared itself publically to be a nuclear-weapons state. Israel comes immediately to mind. Other situations apply or could apply. India, based on its original nuclear explosion, had a virtual arsenal for more than two decades. Threshold nations certainly have had the capability to rapidly manufacture a few nuclear weapons, and for all we know, may actually have some stored away (as South Africa did).

Needless to say, in order for deep reductions by the main nuclear-weapons powers to take hold, widespread surveillance and/or inspections (that include threshold states) would have to take place. For disarmament to take place, not just the acknowledged weapon states would have to join in commitments and openness.

Strategic Reviews

Every four years the United States military has gone through a Congressionally mandated major review of strategy; these quadrennial defense assessments take place at the beginning of each presidential administration. The exercises normally lead to presidential directives on national security.

In 1992, under President G.H. Bush, the first of the post-Cold-War reviews took place. It came up with the concept of a "Base Force" that trimmed military strength without major realignment. The Clinton Administration's "Bottom-Up Review," completed in 1993, simply piled new modes of operation on top of old preparations. A Congressionally assigned 1997 review panel criticized that quadrennial review because it continued to allocate significant resources to the "low-probability" scenario of fighting two major wars at once.

These commissions usually fail to spark deep restructuring because they rely on military-service evaluations (that devise *military* strategies), rather than executive-branch guidance (for overall *national-security* strategies). As a case in point, Clinton's Defense Department continued to assume "forward deployment" of U.S. forces in foreign countries — retaining the existing division of "roles and missions" among the military services while limiting reliance on capabilities of allies. Simply setting budgets limits is not enough to encourage new military strategies and roles.

Soon after taking office, President G.W. Bush was expected to respond in six months to a call from Congress for a new national-security strategy. A Commission on National Security was simultaneously created by Congress to take a longer-range view (out to 2025).

In any event, the military services have a parochial dilemma: choosing between spending money on nuclear or conventional weapons. In that respect, a retired military officer — who became a nuclear-defense analyst — remarked that "Nukes are not considered a usable viable weapon ... anymore." Conventional forces were being recast as the third leg of a new strategic triad (with nuclear missiles as the first leg and missile defenses as the second). A Los Alamos political scientist acknowledged that conventional forces "will play a greater part in deterrence in the future," recognizing that nuclear weapons have transitioned into a "silent role."

Within the context of strategic security, conventional national-defense forces and alliances has become a dominating factor. NATO — strengthening conventional-force retaliation, shifting away from nuclear responses, and expanding membership — has continued to be the major regional alliance for a large partnership of Western nations. To avoid political irritations, NATO has had to take into consideration regional interests of the CIS alliance, which includes nations that were once part of the Soviet bloc.

PDD 60. Approved by Clinton in 1997, Presidential Decision Directive 60 evidently retained U.S. dependence on a large arsenal of nuclear weapons. Although PDD 60 was classified as a secret directive, some of its provisions have been disclosed. The strategic triad was retained, launch-on-warning remained part of the deterrent posture, the right of first-use of nuclear weapons was asserted (including against non-nuclear-weapon states), and targeting plans were extended beyond Russia and China to embrace prospective nuclear states. Had significant strategic reductions been anticipated, some long-range preparations justified by the directive would not have been needed.

U.S. nuclear doctrine during the Cold War had two primary objectives: deterring nuclear attacks and limiting damage if deterrence failed. Held at risk were valued Soviet political and military assets, including nuclear forces, command and control, conventional military forces, the political and military leadership, and their defense industries. The strategy was chosen because defense analysts thought that "implacable, aggressive, and risk-taking [Soviet] adversaries" would place greater value on "the perpetuation, expansion, and aggrandizement of their regime" than they would on "the lives of ordinary citizens." Eventually the analysts learned they were wrong about that.

Anyway, damage limitation (in case deterrence failed) could have been achieved by limiting the scope of nuclear attacks, holding Soviet cities as hostage to nuclear attack, and degrading Soviet capacity to execute nuclear attacks. That was the old strategy of "countervalue," a doctrine violating international law which forbids threats or use of force directed against civilians.

Continuation of classic nuclear strategy seems to have been in the minds of military and civilian nuclear strategists who influenced the 1994 Nuclear Posture Review. Carried over from the bilateral confrontation, impacting U.S. domestic perceptions, was an unwillingness to relinquish national-security dependence on large and active nuclear forces.

Although Clinton's PDD 60 cleared the way for cuts in strategic warheads and reportedly "eliminated the requirement to prevail in a protracted nuclear war," it "nevertheless maintained the basic structure of U.S. [Cold War] nuclear doctrine and targeting plans." The directive called for U.S. nuclear forces to "prevail even under the condition of a prolonged war." This philosophy prevailed despite recognition of risks of "accidental, unauthorized, or erroneous attacks triggered by a breakdown in the Russian command-and-control system."

As a result, authors of *The Nuclear Turning Point* remarked that "even an *option* to launch under attack is unwise because it forces political and military leaders to make momentous decisions in a few minutes with incomplete information on the nature or origin of the [presumed] attack." To offset this risk, they suggested the United States should take the lead in announcing that the only purpose of nuclear weapons is to deter attacks — thus adopt a no-first-use policy and renounce counterforce targeting.

Other alternatives to counterforce targeting include "counterpower targeting" (aiming at conventional military forces and defense industries) and "adaptive targeting" (a flexible response not tied to a fixed nuclear-war plan). As horrible as it might seem, countervalue targeting — threatening to destroy cities and industry — might be more palatable. It translates into a policy of minimum deterrence, meeting McNamara's "assured destruction" criterion. Some precedence already exists in that Britain, France, and China long ago settled on a minimum-deterrence posture.

Contingency planning for minimum deterrence could be guided by several assumptions: no-first-use of nuclear weapons, no hasty response with nuclear weapons, and no retaliation that is not tailored to the circumstances. Under these conditions, 10 to 100 survivable nuclear weapons would be adequate.

Taken in accordance with or legitimated by PDD 60, potential nuclear weapon enhancements tend to be inconsistent with minimum deterrence. Counterproductive steps include advanced nuclear-warhead modeling procedures supported by sub-critical explosive testing and use of the National Ignition Facility for weapons upgrades.

Other steps detrimental to minimum deterrence under recent U.S. policy directives include arrangements to resume tritium production (using civilian power reactors) and continued (limited) production of plutonium pits. Both of these measures would be superfluous if nuclear reductions were negotiated. Because the Defense Department, in its 1994 Nuclear Posture Review, adopted a "go-slow" approach, nuclear reductions have been shunted aside. By omitting initiatives for carrying out or negotiating significant strategic cuts, the slow track was formalized in PDD 60.

Having interviewed many key personalities of the arms-control scene, Jonathan Schell concluded,

> Deterrence is an abstract doctrine, yet it risks the most drastic conceivable material consequences. It's a product of pure thought whose centerpiece, nuclear annihilation, is commonly called "unthinkable." Seen in one light, it's a doctrine of limitless intricacy, shot through with conundrum and paradox.... It's a system of terror that purports to hold terror at bay.

A stumbling block in worldwide nuclear elimination, pointed out by Professor Bruce Larkin in a detailed study of British, French and Chinese weapons, is that these three nations (and Russia) are highly dependent on the nuclear deterrent, not only to counter a nuclear attack, but also to defend against massive conventional attack. As a result, Larkin concluded that a "disengagement regime" — for which he meant that nuclear weapons would be retained, but without prompt access — would be better than abolition. Larkin thought that there were some "net good" aspects of nuclear deterrence, including prevention of conventional war, guaranteed existence of state and society, and enablement of political actions. Larkin was pessimistic about abolition for several reasons: inability to prove that nuclear-weapons states will not hold back some weapons, no way to prevent with certainty new nations from making nuclear weapons, and the possible emergence of other weapons of mass destruction.

Bush's NPR. The essence of the G.W. Bush administration's first nuclear-posture review was widely disseminated (or deliberately leaked), despite its security classification level. The 2001 NPR outlined "a conceptual path" toward an operationally deployed force of 1700-2200 warheads in 2012, and suggested the number of operationally deployed warheads could be subsequently reduced below 3800.

According to the *Chicago Tribune*, the NPR recommended "trimming the nuclear fleet of Trident submarines from 18 to 14, no longer using the B-1 bomber as a nuclear-arms platform and destroying 50 Peacekeeper ICBM silos...."

But the *Tribune* called the study "disappointingly modest" because it lacked "bold ideas for deep, unilateral reductions in strategic warheads that Bush has promoted since his 2000 campaign." Other reported disappointments are that "The U.S. would put itself in position to resume nuclear testing," "some 2200 sites in Russia would remain nuclear targets, a throwback to Cold War thinking," "strategic arms cuts would be stretched out over 10 years," and "many of the deactivated weapons would be kept in storage rather than destroyed."

The *Chicago Tribune* also noted that if U.S. nuclear weapons are simply kept in storage, "Russia will have little incentive to build down its nuclear arsenal in tandem with the United States. The greater risk is that "Russian nuclear material will fall into unfriendly hands." With regard to nuclear testing, "a signal that the United States is preparing to resume such tests would encourage other nations to do so as well. That would be a grave mistake," added the Republican-leaning newspaper.

Although mutual assured destruction remains the *de-facto* condition, the Bush administration sought more alternatives in using nuclear or non-nuclear responses to perceived threats. Greater emphasis on conventional and defensive capabilities was intended.

One of the more disappointing aspects of the Bush administration's review was retention of nuclear counterforce as a strategic doctrine. Defense Department initiatives for U.S. military forces were centered on nuclear-centric goals: holding at risk critical mobile and relocatable targets; defeating hard and deeply buried targets; developing capabilities for long-range strikes in all terrain and weather conditions; converting submarines from ballistic-missile to guided-missile service; and enhancing nuclear-strike capabilities for timely arrival on target, improved precision, and rapid re-targeting.

The FAS believed that "Unless counterforce is abandoned, we cannot honestly say that we have broken with the Cold War." The respected NGO challenged the morality of a counterforce strategy because an attack on Russia's nuclear installations "would still kill tens of millions" in its adjacent cities. Moreover, the FAS feared that targeting Russian missiles directly caused Russia to maintain its missiles on hair-trigger alert, thus prolonging "the real and terrifying possibility of an accidental nuclear attack on America."

In examining the nuclear posture review, the *Chicago Tribune*, was pleased with what the war on Afghanistan demonstrated and satisfied that "U.S. conventional arms are now so sophisticated and fearsome, they could deter a terrorist or rogue state from attacking the U.S. at least as effectively as the threat of our nuclear retaliation." However, a dissident member of the *Tribune's* editorial staff, Steve Chapman, deemed the new Bush nuclear policy to have "old flaws" that make the world "*less safe*." Chapman came to this conclusion by observing that the proposed cuts "aren't really cuts," because many of the warheads would not be destroyed but "simply put in storage for some rainy day." The problem with that, he noted, was that "the Russians will do the same thing," keeping warheads and nuclear materials that "might find their way into the hands of terrorists."

Chapman asked, "why do we need to be able to incinerate [Russia] in less time than it takes to watch the evening news?" He cited that as an illustration of clinging to "the old way of doing things" while the Bush administration was boasting that it has "put Cold War thinking behind." Chapman also took issue with the administration's desire to pour money into nuclear test preparations, pointing out that

> No country in the world has less reason to conduct nuclear tests than the U.S.... A ban on all tests would serve the selfish purpose of preserving American nuclear superiority, in quality as well as quantity.

On behalf of the 2001 NPR, the Bush administration declared that

> U.S. forces are not on "hair-trigger" alert, and rigorous safeguards exist to ensure the highest levels of nuclear weapons safety, security, reliability, and command and control. Multiple, stringent procedural and technical safeguards are in place to guard against U.S. accidental and unauthorized launch.

While these assurances might very well be valid, many outsiders found that their own understanding of the U.S. nuclear-alert status to be less comforting than that of the Defense Department.

[America's Cold War Nuclear-Weapon Obsession (Bruce Blair, CDI, 2005)]. Much clamor is being stirred by the Pentagon's plans to develop specially designed nukes to ... preemptively ... neutralize the emerging rogue nuclear threats.... [Neither] the perception of a rogue nuclear threat nor the idea of resorting to U.S. nuclear weapons to suppress it is new to Pentagon planners. On the contrary, a nascent nuclear threat was attributed to these very same countries over 20 years ago, and nuclear strike plans were devised to suppress it....

War-gamers argued that Iran or one of the other putative nuclear rogues ... or China ... might emerge from the ashes of a U.S.-Soviet nuclear exchange and exploit U.S. weakness using nuclear blackmail.... U.S. planners simply assumed the extreme worst-case for both the capabilities and the intentions of the inscrutable and angry regimes in Iran and elsewhere.... Now that the U.S.-Soviet nuclear rivalry has become the side-show, and nuclear proliferators have stolen the limelight, U.S. nuclear planners enjoy new license to conceive scenarios for using U.S. nukes against the rogue states and China....

The war-gamers lost their credibility and perspective on the utility of U.S. nukes in dealing with nuclear rogue states over two decades ago. They are still living in that strange dreamland....

This discussion of Strategic Reviews has not been joined with tactical nuclear policy partly because it is taken into account under the preceding subtopic of Strategic Deterrence. Post-Cold War nuclear operations have more to do with perceived regional political stakes than they do with strategic deterrence. Theater nuclear goals extend to any part of the world where it might be necessary to dissuade adversaries "from undertaking programs or operations that could threaten U.S. interests or those of our allies and friends." Moreover,

> like any military action, **the decision to use nuclear weapons is driven by the political objective sought** [emphasis in original]. This choice involves many political

considerations, all of which impact nuclear weapon use, the types and number of weapons used, and method of employment.

Because theater nuclear policy is deliberately activistic in promotion of national political interests, contemporary strategic policy is relegated to being a passive factor in sustaining national security.

Risk-Reduction Measures

Pending a massive change of heart by nuclear-weapons states, several interim risk-reduction measures en-route to deep cuts have been proposed. Some are mentioned in this subsection, with elaboration on others in subsequent parts of this Chapter.

Joseph Rotblat advocated a nuclear no-first-use treaty as an initial step, arguing that nuclear weapons should be assigned the single role of deterring other nuclear weapons. Robert McNamara, who was Secretary of Defense during the Caribbean missile standoff, belatedly became a champion of no-first-use.

Helmut Schmidt, former Chancellor of West Germany once called for deployment of Pershing missiles. Turning in favor of abolishing all nuclear weapons, Schmidt recognized that getting rid of them would not occur until everyone else was doing the same. He noted that while "Americans now think they are the only superpower in the world," this is a misunderstanding: Russia is "still a [nuclear] superpower."

Jonathan Schell, in the "Epilogue" to his article in *The Nation* on "The Gift of Time," mentioned that

> Political will ... can be given institutional expression. A no-first-use policy, for instance, is a declaration of intention that would have both technical and legal consequences. The technical consequences would be a shift in the composition of military forces to provide defense without using nuclear weapons first. The legal consequences would be adherence to a treaty agreeing to no-first-use. However, the great transforming step along the path to political zero would certainly be the renunciation of the retaliatory strike — no second use — for only this step would render renunciation of nuclear weapons complete.

De-alerting — that is, changes of policy and procedures so as to tighten the trigger on ballistic-missile launch procedures — could be carried out without verification because it is in the self-interest of each of the nuclear-armed nations to render "safe" their missile launchers. In 2003, Steve Chapman of the *Chicago Tribune* criticized the need for continued hair-trigger missile-alert postures, noting that the United States could turn an "entire territory into radioactive rubble," and "whether that takes half an hour or half a day could not possibly matter less." He called for "de-alerting missiles on both sides, in a verifiable way," to "create a cushion of hours or even days, so that neither side can take a hasty and ill-informed leap in nuclear war." Chapman agreed with John Steinbruner that it is "the single most important thing we could do, in any area, to improve the security of the United States."

Senator Alan Cranston was quoted by Schell as saying, "When we get down near to where the UK, France and China are [in nuclear arsenals], we need to draw them in, and move together to very low numbers. At that stage, we should draw in the threshold states — India, Pakistan and Israel."

To decrease the tacit nuclear confrontation that still exists between the United States and Russia, trust-building and restraint-exercising are fundamental steps. Because of limitations in existing dismantlement facilities, destruction of strategic and tactical nuclear weapons can only be gradual. In the meantime, technical and institutional demilitarization procedures are being developed, and decisive steps are being implemented and/or evaluated for weapons-grade uranium and plutonium.

The preceding are just a few of the more substantive measures that can be undertaken while the world waits for the nuclear-weapons states to reach agreement on major reductions. In addition, unilateral declarations can set the tone for verifiable reductions.

Declaratory Policies. The Bush-administration declaratory policy quoted above left little doubt about using nuclear weapons whenever seen fit, especially when coupled with advertised forward-basing of nuclear weapons.

For those who propose that reductions in nuclear risk proceed first through a stage of less-bellicose declaratory policy, one can easily see the difficulties involved in taking comfort simply from public statements; for those who prefer parallel retrogressive stages, moderation of declaratory policy would be welcome as one of many steps in backing down from the nuclear brink.

Policies that depend on government declarations are important but largely symbolic. Examples are no-first-use, de-alerting, and de-mating of nuclear-weapon systems. They are useful steps toward more enduring measures of actual disarmament. However, unless declaratory policies are followed by supplementary measures that include mutual verification in order to impede rapid unilateral breakout, the pledges would have little military significance. This is a point hammered home by French arms-control analyst Thèrése Delpech, who examined military records of the Warsaw Pact: These "demonstrated without doubt that [Soviet] operational plans called for the use of nuclear and chemical weapons in West Germany at the onset of hostilities [even if NATO forces were using only conventional weapons] — this at a time when the [Soviet] official doctrine was no-first-use."

Official no-first-use declarations — asserting that the sole purpose of nuclear weapons is to deter and, if necessary, respond to the use of nuclear weapons — would indeed enhance national security and promote nonproliferation. But such declarations would become more plausible if accompanied by meaningful reductions in nuclear forces and safer nuclear postures.

Arms Control vs. Nonproliferation

Distinguishing between arms control and nonproliferation is increasingly important; the objectives are not necessarily the same, and often one is affected by the other. Nuclear arms control is directed at confining and eliminating excess inventories of weapons and their components (reversing vertical proliferation); nonproliferation is aimed at deterring indigenous production of weapons in non-nuclear-weapons states (thwarting horizontal proliferation).

Thus, maintaining nuclear arsenals and preventing proliferation are nominally incompatible objectives of nuclear-weapons states, particularly the United States, which encouraged and tolerated nuclear weaponization by its allies. Weapon-states engaged in their own nuclear armament with little real regard to its effect on nonproliferation. Since then, stagnated policies maintaining excessive arsenals, production capability, and nuclear materials have perpetuated the predicament.

Domestic squabbles over policy direction have also impeded progress in arms control. For instance, indiscriminate opposition to nuclear reactors has provoked unfounded fear of plutonium, whether reactor- or weapons-grade. As a result, interventionist concerns have stymied progress in eliminating weapons plutonium, which would be a major step in irreversible weapons destruction. In fact, reactor-burnup of weapons plutonium could already have been well underway.

To simultaneously promote goals of non-proliferation, nuclear safety and environmental well-being, scientists have encouraged the application of constructive nuclear safeguards. This includes fissile-materials confinement, reactor security, remote monitoring, and fissile accounting. However, it would be wasteful and self-defeating if every variety of plutonium were regarded as having the same explosive potential as high-purity (weapons-grade) plutonium; this, however, has been the nearly implacable stance taken by some anti-nuclear-power obdurates. Even though first-cycle discharged reactor fuel usually contains separable plutonium to hypothetically make a crude nuclear explosive, its speculative diversion would be unlikely to go undetected, nor would it yield a desirable product of military value.

Fortunately, knowledgeable scientists, engineers, and policymakers agree that safeguards and disposition should be allocated to fissile materials in proportion to the danger they represent, allowing for practical limitations.

Absent a program for deliberately sorting plutonium by its quality, demilitarization of weapons-grade plutonium would be indefinitely deferred, keeping it around in easily reusable form for many decades. A better alternative would be to degrade military plutonium in commercial nuclear reactors, while thoroughly safeguarding all separated power-reactor plutonium.

Placing too much significance (as some influential academics have done) on the explosive potential of low-grade, reactor plutonium has had other downsides: The misplaced emphasis has helped scuttle fissile-production cutoff negotiations, endangered renewal of the Non-Proliferation Treaty, and discouraged future arms-control treaties aimed at elimination of nuclear weapons.

Recently, a coalition of three leading arms-control NGOs has publically acknowledged that the continuing risk of nuclear self-destruction from existing nuclear weapons is a more immediate danger than the long-term spread of nuclear proliferation. Perhaps this will lead to broader concurrence with practical steps toward demilitarization of weapons-grade fissile materials.

Arms Control vs. Missile Defense

Missile defenses cannot be developed without impacting nuclear arms control, and arms reductions cannot make substantial headway without factoring in defenses against long-range missile attack. Despite this coupling, each of these objectives continues to be separately energized by its own dynamic.

As the number of nuclear-armed intercontinental ballistic missiles is reduced, the value of building a missile defense is increased; there are fewer potential incoming missiles to counter. A political environment in which strategic reductions are taking place implies an atmosphere conducive to negotiations and agreements; this would provide a firm basis for a gradual buildup of missile defenses that could become technologically feasible.

U.S. right-wing pressure for a national missile defense has been persistent, as well as vocal, especially prior to and after President Bush rescinded the ABM Treaty. Funds allocated by Congress during the Bush administration drove a premature, unrealistic, and wasteful schedule for deployment.

A rare leak of negotiating documents during the Clinton administration revealed the original "talking points" on the ABM Treaty while it was still being negotiated with Russia. The U.S. "approach" for verification was based on four fundamental "elements": information exchange, notification of key events, inspections, and resolving matters of concern. The Clinton team also hoped to reach simultaneous agreement on a Russian proposal for development of a global monitoring system and cooperation on missile tracking and proliferation of missile technology. At the time, the administration looked favorably on global monitoring of missile launches, which would include notification, exchange of early warning information, and establishment of an international coordinating center.

Unfortunately, the Clinton "talking points" revealed that, in order to achieve their ABM goals, U.S. negotiators were willing to sacrifice arms control: ruling out future reductions in strategic nuclear warheads below the 1500 to 2000 level by rejecting an offer to pare down then-current levels of 6000 to 7000 warheads. The administration was also, in effect, encouraging Russia to maintain strategic nuclear forces on hair-trigger alert, ready to fire within minutes of receiving a launch order.

The Nuclear Turning Point argued that "unless a fundamental change in deterrence relationships occurs *before* the deployment of significant levels of strategic-capable defenses, their deployment will almost certainly prevent the attainment of ... deep reductions...." Yet, the case can be made (as it is in *Nuclear Insights*) that cooperative reductions could and should go hand-in-hand with a defensive buildup.

U.S. missile defense is no longer primarily concerned with a large-scale, deliberate missile attack from the former Soviet Union. The threat from Russia has diminished considerably. Bruce Blair, head of the Center for Defense Information, held the opinion that "To overwhelm an NMD shield, Russia must plan to launch massively and quickly in a crisis, either firing first or firing on warning from a deteriorating network of early warning satellites of an incoming missile strike." Missile-defense advocates began to stress deliberate attacks from so-called "rogue" nations, as well as accidental, unauthorized missile launches.

The Nuclear Turning Point detailed potential threats from short-range, medium-range, and intercontinental ballistic missiles. One the other hand, *Nuclear Insights* focuses primarily on the long-range, nuclear-armed missile menace. A 1995 U.S. national-intelligence report estimated that it would take up to 15 years before new nuclear powers could develop an intercontinental capability. That estimated time projection has been both supported ("Gates" review panel) and challenged ("Rumsfeld" Commission).

Absent a strategic confrontation, the primary residual threat to any nation, especially former adversaries, is the accidental or unauthorized launch of nuclear-armed RVs from thousands of existing ICBMs. Although this has very low probability, it would have extremely devastating consequences. The danger is greatest for sold-fueled ICBMs topped with multiple RVs. Nations such as China that have liquid-fueled missiles can and usually do keep them unfueled in order to minimize such dangers. Now that the Cold War face-off has subsided, an additional safety and security measure would be to dismount all warheads, placing them on standby in close proximity.

Under pressure from Congress, the Clinton administration had developed a three-year plan to move ballistic-missile defense from a technology-readiness program to a deployment-readiness program. But developers were unable to respond on such a short time scale to deploy a "thin" NMD. In any event, achievement of nationwide coverage would have required sensors at multiple locations or in space to obtain the needed tracking information, and such deployment was banned by the ABM Treaty.

Theater missile defense, not banned by the ABM Treaty, was vigorously pursued by the Clinton administration. Programs included an upgraded Patriot interceptor, two modified naval air-defense systems, a joint American-European troop-protection project, the THAAD Army missile-defense system, and an airborne laser. The latter are attempts to intercept theater ballistic missiles in their boost phase, which is before the payload is released. At the 1997 Clinton-Yeltsin summit in Helsinki, an acceptable demarcation for theater defense was signed and confidence-building measures were put into effect.

One big problem with most existing missile-defense schemes is that countermeasures are relatively simple and have already been implemented. For example, in the vacuum of space, warheads and decoys travel on parallel trajectories in such a manner as to offer too many targets to an interceptor. Decoys can be cleverly camouflaged to confuse target acquisition, and missiles can be maneuvered into evasive trajectories. A boost-phase defensive concept proposed in Section G of this book does not have that particular weakness to countermeasures.

Another major problem with NMD is circumvention, that is, using means other than ballistic missiles for delivery of nuclear warheads; e.g., nuclear-armed cruise missiles can be launched off-shore from submarines or ships. Methods of ground, air, and sea delivery exist that will not leave a clear "return address" for retaliation.

The Nuclear Turning Point reckoned that most missile-defense proponents tend to inflate estimates of capabilities "because their effectiveness is difficult to assess reliably in advance and because attackers and defenders will form very different assessments." Those authors feared a "defense breakout" if offensive forces are significantly reduced. (However, they considered too briefly coordinating force reductions with defense buildup.)

One way to diminish the threat from ballistic-missiles, according to *The Nuclear Turning Point*, would be for the major nuclear powers to "eliminate ballistic missiles as strategic nuclear-delivery systems in favor of bombers, cruise missiles, or some other type of delivery system (or do away with nuclear weapons altogether)."

Another way is to restrict "the proliferation of ballistic and cruise missiles, unpiloted air vehicles, and related technologies." For this purpose, the Missile Technology Control Regime was established in 1987; it is not a formal treaty "but a voluntary agreement among a group of [29] countries to apply export controls to a common list of complete missiles, missile subsystems and components, and missile-related technologies." Some other countries, including China, are not members but have agreed to adhere to its guidelines.

As expounded by George W. Bush when he was a presidential candidate in June 2000, the Republican Party position called for an all-out national missile defense accompanied by unilateral reductions in nuclear warheads. The proposed number of reductions was not specific, nor did Bush endorse a doctrine of "minimal deterrence" (the lowest number of offensive nuclear weapons necessary to inflict unacceptable damage on an attacker). Coupling both reduced weapons and increased defense is desirable, but the manner and timing of such a policy would be important. Premature deployment of a missile defense could undermine confidence in mutual deterrence.

The G.W. Bush administration's approach was likened to "ahead full speed, damn the torpedoes." On the eve of a fourth test of kinetic-missile intercept, the Pentagon's Deputy Defense Secretary, Paul Wolfowitz, told the Senate Armed Services Committee that "We have to withdraw from [the ABM Treaty] or replace it." At the same time, National Security Advisor Condoleezza Rice claimed on behalf of the administration that "We look to move cooperatively with the Russians" asking rhetorically, "Why, when we are not enemies, are we still postured as if we were in the Cold War?" Apparently she was not then in tune with her own political party.

In response to the Bush-administration plan to scuttle the ABM Treaty, a Democratic Senator, Jack Reed, charged that "The real [Republican] agenda appears to be replacing arms control with an arms buildup. I believe what's happening is there is a conscious rejection of arms control as a central tenet of American foreign policy."

A (multi-partite) negotiated missile defense, as described later in this book, elements of which are contained in a new Obama administration plan (see box) could be negotiated. Under circumstances that would encourage development of an agreed defense, many of the present-day technical, military, and political difficulties in missile-defense deployment would then be avoided. European governments, which considered the ABM treaty an essential element of international security, clearly preferred to ease Russia's anxiety by supporting a negotiated amendment process.

[President Obama Modifies Missile Defense Plan]. On 17 September 2009 the following statement was issued by the White House:

The best way to responsibly advance our security and the security of our allies is to deploy a missile defense system that best responds to the threats that we face, and that utilizes technology that is both proven and cost effective.

[This] new ballistic missile defense program will best address the threat posed by Iran's ongoing ballistic missile defense program [short and medium-range missiles, which are capable of reaching Europe].

Our new approach will therefore deploy technologies [land- and sea-based interceptors and the sensors] that are proven and cost effective and that counter the current threat... [It will be phased and adaptive], and we will retain the flexibility to adjust and enhance our defenses as the threat and technology continue to evolve.

This approach is also consistent with NATO missile — NATO's missile defense efforts and provides opportunities for enhanced international collaboration going forward.

[We] welcome Russians' cooperation to bring its missile defense capabilities into a broader defense of our common strategic interests, even as we continue to — we continue our shared efforts to end Iran's illicit nuclear program.

Transition to better mutual security could be initiated through collaboration on missile (and aircraft) surveillance, which is a central feature of any missile defense. Proposals for joint U.S. and RF ballistic-defense research and development are not unreasonable. Russia (and China) could be integrated into the sensing and information-management network that is needed to support any deployment. Comprehensive collaboration in monitoring missile launches and air traffic would also be valuable steps in creating a reassuring and stable deterrent environment. The collaboration could be extended to the pre-launch conditions associated with all nuclear-delivery systems. If such negotiated conditions for decisive and concurrent reassurance could be reached, missile-defense deployment would not aggravate doubts about strategic stability.

The Obama regional missile-defense plan is to unfold in three stages. By 2011, the Pentagon will deploy Navy Aegis ships equipped with SM-3 interceptors in the eastern Mediterranean that are linked with sensors, giving the military the ability to defend critical infrastructure and U.S. forces in Europe.

A second phase in about 2015 will field an upgraded, land-based SM-3 in allied countries, including Poland and the Czech Republic. The upgraded SM-3 will be coupled with airborne sensors that will expand the covered area threefold.

In 2018, the third phase will deploy a larger and more capable version of the SM-3 which will allow the missile defense shield "to cover the entire land mass of Europe against intermediate and short-range ballistic missiles." Perhaps by 2020 the system will be made more powerful to intercept intercontinental ballistic missiles (ICBMs) fired from Iran at not only Europe but the United States.

Three Aegis ships, each carrying about 100 SM-3s, would be deployed at any given time in and around the Mediterranean and the North Sea to protect "areas of interest." The program would also involve the deployment of a mobile, X-band radar in the Caucasus — the same type of radar currently used in Japan and Israel.

The Pentagon seeks to involve Russia in the new system in part by integrating radar systems in southern Russia into the network to give "greater coverage to potential Iranian missile launches."

As long as ballistic-missile defense deployment doesn't get ahead of its technology or the ambient strategic environment, defense augmentation carried out in phase with negotiations on nuclear-arms reductions would make a win-win state of affairs for all concerned.

Outsider Opinions

Non-government individuals and organizations have expressed views often highly incompatible with governmental positions. Yet, NGOs had an appreciable influence in stimulating restraints over excessive nuclear arsenals and hawkish war-advocacy policies.

The traditional political and diplomatic process, left to itself, does not seem sufficiently motivated or effective in promoting nuclear reductions or disarmament. Controlling the military-industrial-laboratory-congressional quadrangle has required, and continues to require, external oversight. A sampling of outsider views and activities is provided below.

When a concerned Pakistani citizen was asked about his nation becoming a nuclear-weapons state, he responded:

> [If] nuclear deterrence guarantees peace, why is defense spending by [India and Pakistan] continuing to rise? And if nuclear weapons are "the cheapest means of defense," why are the budgets for conventional forces increasing?

Bruce Blair, then at the Brookings Institution, is quoted in 1998 by Jonathan Schell as saying "No major change in the U.S.-Russian nuclear equation has occurred — not in war-planning, not in daily alert practices, not in strategic arms control, and maybe not even in core attitudes." However outmoded, the doctrine of massive deterrence remained in place more than a decade after the collapse of the USSR.

Blair specifically proposed de-MIRVing missiles: If the multiple warheads topping 500 Minuteman-III missiles were reduced to one, it would cut 1000 nuclear weapons; if ten on 50 MX were reduced to one, it would slice another 450. Thus, the land-based U.S. arsenal could go from 1500 to 550, with the bonus that each missile would no longer be a prime target in a pre-emptive, counterforce strike. The warheads could initially be put in storage and eventually become part of an agreement that enabled their dismantlement.

[Restructuring U.S. Strategic Forces]. Sidney Drell in 2007 suggested a "fresh look" at the role of nuclear weapons in U.S. defense planning.

The United States and Russia have now officially adopted a policy of cooperation against new threats, faced by both nations, of terrorists and unstable or irresponsible governments acquiring nuclear weapons. This replaces the former adversarial relationship of nuclear deterrence based on mutual assured destruction.

The goal of the revised force structure is 500 "operationally deployed" nuclear warheads, plus 500 in a "responsive force," a drawdown to take place in five years.

The "operationally deployed" force would consist of three Trident submarines at sea, each carrying a total of 96 warheads; 100 Minuteman III ICBMs in hardened silos, each with a single warhead; and the balance of warheads carried by B2 and B52H bombers "configured for gravity bombs or air-launched cruise missiles."

The "responsive force" would consist of three more Trident submarines that are either in transit or being replenished in port; 50 to 100 additional Minuteman III missiles without warheads; and 20-25 unarmed bombers in maintenance or training.

Because of the large number of residual tactical nuclear warheads, arms reductions would be myopic if limited to strategic missiles. The United States has about 1000 tactical warheads in active service (and perhaps 3000 in reserve and storage). Russia is said to have about the same number of non-strategic warheads, although the Pentagon has claimed Russia retained more than 15,000. Fortunately, contemporary Russia is anxious to reduce the burden of safeguarding and maintaining their tactical nuclear weapons.

Panofsky. One prominent outsider involved in the military-policy debate was Wolfgang Panofsky (often consulted by presidents), who reasoned:

There are three overriding lessons to be learned from the dangers inherent in the excessive production of lethal devices during the Cold War and today's difficulties in dealing with the resulting surplus. First, physical realities must be communicated honestly to decision-makers and must not be overridden by politics or perception. Secondly, science and technology cannot be coerced by political mandate to deliver what cannot be achieved. Finally, an old adage, *Si vis pacem, para bellum* (If you wish peace, prepare for war) should be changed to *Si paras bellum, utque para pacem* (If you prepare for war, also prepare for peace)!

Because of Panofsky's longtime official and unofficial involvement in these issues, his analysis on the role of nuclear weapons has been worthy of particular attention:

The Cold War is over but little has changed with respect to U.S. nuclear weapons policy. Yet the nature of the threats to U.S. security from nuclear weapons has shifted dramatically since the end of World War II. Today the likelihood of deliberate large-scale nuclear attack against the U.S. is much less than the risk of a nuclear weapon accident, unauthorized use, or the threat from the proliferation of nuclear weapons across the globe.

Observing that the nuclear buildup peaked at more than 60,000, and the builddown had not yet reduced that inventory by a factor of two, Panofsky took issue with the 1994 U.S. Nuclear Posture Review that retained "a great deal of Cold War thinking." Since then, the only revision in policy has been that the United States need no longer be prepared to fight a "protracted" nuclear war.

In Panofsky's opinion, the United States has the most to lose if nuclear weapons proliferate to more nations.

CISAC. In a 1997 CISAC report for the U.S. National Academy of Sciences, "prohibition" of nuclear weapons was favored. The three benefits identified were (1) almost complete elimination of possible use, (2) reduction in likelihood of additional states acquiring nuclear weapons, and (3) decisive resolution of the uncertain moral and legal status of nuclear weapons. The recognized risks of prohibition included (1) a breakdown in stability caused if a nation cheated, (2) overt withdrawal from the disarmament regime, and (3) less moderation in the behavior of some states.

That particular Academy committee endorsed interim measures, including superpower reduction to 1000 warheads, eventually to be followed by reductions to a few hundred on each side. De-alerting was favored, along with removal of key missile components, such as warheads, shrouds, or guidance systems. All reduction steps would be monitored, as well as a comprehensive accounting of warheads and a cutoff in production of fissile materials. Overall, an evolutionary approach was recommended by the committee, which included Wolfgang Panofsky, John Steinbruner, and Maj. Gen. William Burns (former ACDA Director).

INESAP. International networks, such as INESAP, which includes 60 organizations from 25 countries, continued to promoted nuclear disarmament and the tightening of arms control and non-proliferation regimes. The network is comprised primarily of scientists and engineers who support INESAP activities that speak out about nuclear testing, ballistic-missile testing and defense, fissile-material production, and reductions leading to elimination of nuclear weapons.

Harald Müller, director of the Peace Research Institute Frankfurt and a well-known nuclear abolitionist, is quoted as saying:

> There is a sickness in the American mind which says that everybody really wants to have the bomb because we have it and we don't want to get rid of it. But ... consider a country like Belgium, for instance. It has plutonium. But this does not mean it has a "virtual arsenal." It's so silly ... because there is no thinking about nuclear weapons in any Belgium mind. They used to think about them, but now it's gone. We have to beat back this sick culture that tells us that everybody wants the bomb.

FAS/NRDC/UCS. An appropriate nuclear posture in the new millennium, according to an ad-hoc alliance of NGOs, should contain the following nine particulars:

(1) declaration that the sole purpose of nuclear weapons is to deter or respond to nuclear attack;
(2) rejection of rapid-launch options and deployment practices;
(3) development of tailored responses for targeting plans;
(4) reduction in U.S. (and Russian) nuclear arsenals to about 1000 warheads, including deployed, spare and reserve;
(5) retirement and dismantlement of tactical weapons;
(6) commitment to further reductions on a negotiated and verified multilateral basis;
(7) pledges not to resume nuclear testing, while ratifying the CTBT;
(8) reaffirmation of the pursuit of nuclear disarmament; and
(9) recognition of the implications of missile-defense deployment.

The NRDC, independently updating its nuclear-data books, has created a much greater awareness of the status and magnitude of nuclear arsenals. Both the FAS and UCS continue to carry out their own respective programs of analysis and activism to reduce nuclear risks, making good use of the Internet to improve networking with other organizations.

To sustain activism on nuclear issues, the UCS organized ArmsNet, a grass-roots engagement with government officials and legislators. Scientists and other informed individuals were brought to Washington, D.C., on "Education Days" in order to inform and influence policymakers about arms-control issues such as missile defense, conflicting priorities in defense spending vs. combating terrorism, and the robust nuclear earth penetrator.

The (G.W.) Bush Presidency

As President G.W. Bush's term got underway, opportunities existed to make progress not achieved during earlier stages of the Cold War phaseout. President Clinton did not concentrate on arms-control issues, and he was hampered by a partisan Congress. President Bush initially indicated that he preferred some unilateral measures: de-alerting of nuclear forces, reducing nuclear weapons (as his father achieved), and continuing a moratorium on testing. He spoke in favor of inventorying nuclear materials and asking the Republican Congress to fund the dismantlement of Russia's weapons.

Although unilateralism was a major tenet of Bush presidential strategy, it was quickly junked. Defeating the Taliban and Al Qaeda in Afghanistan required assistance from many countries, and interrupting terrorist infrastructure around the world required cooperation of numerous nations. The administration's advertised contempt for signed treaties was partially weakened when President Bush was pressured into signing a written (albeit minimal) nuclear arms-control treaty (SORT) with President Putin.

Partisanship drove funding for a proposed comprehensive missile-defense system with land, sea and/or space-based interceptors. Yet, enormous technical and fiscal hurdles had to be overcome in order to deploy an effective defense.

Although the ABM Treaty was cast aside, President Bush reluctantly understood the value of negotiating with Russia, sharing technology, and consulting with allies.

Emphasis on unilateralism in arms-control is ironic, given that Republican administrations had taken a hard-line view. They demanded strong verification, a process that requires temporarily relinquishing a measure of sovereignty to a treaty signatory or to an international organization. Going it alone has obvious counterproductive consequences.

Because "unverifiability" of the CTBT continued to be a frequent Republican contention, no official verification mechanism was put into practice; yet, a voluntary moratorium on nuclear testing has been in effect. A couple of nations — such as India and Pakistan — have not abided by the test ban's non-proliferation norm because, they say, universal (*e.g.*, CTBT) restraints on nuclear detonations are lacking.

When the United States unilaterally abandoned the ABM Treaty, a former Clinton administration official warned: "Russia may decide to make good on its threats to withdraw from START, the [INF] Treaty, and other international arms-control agreements. Unilateralism in this case would be taken to a destructive edge, far removed from the concept of cooperative action conducted in parallel."

The Clinton administration had looked forward to irreversible, though gradual, reductions in nuclear weapons, but it made little headway. The Bush administration reneged on progress by demanding new accounting rules: disarm by setting the ammo aside, but keep the weapons.

Before and after reaching his one (SORT) agreement with Russia, President Bush announced arms reductions, except that he confined them to "operationally deployed strategic nuclear warheads." It didn't take long for Russian commentators to note that the United States simply wanted to remove strategic-delivery systems and warheads from operational deployment but keep them handy. Peacekeeper (MX) missiles would be removed from their silos, but preserved; they could be redeployed in Minuteman III silos. The Navy would convert four submarines to cruise-missile platforms, but retain the ability to reload ballistic missiles and arm cruise missiles with nuclear warheads. B-1 bombers would be excluded from the counting rules, even though they could be equipped with nuclear bombs that are kept in storage.

Considering that fiscal limitations in Russia motivate elimination of their aging strategic-delivery systems, the United States missed an opportunity to increase overall security and stability.

Ignoring widespread public disagreement, the Bush administration pushed for the "robust" nuclear earth penetrator. Although the immediate effects of a shallow nuclear detonation would probably be localized, radiation from nuclear "bunker-busters" is prone to be much greater and to spread more widely than fallout did after the atomic bombing of Hiroshima and Nagasaki.

In any event, the September 2001 terrorist attacks — followed by unparalleled Bush-administration counterterrorism — diverted international attention from problems of nuclear reductions and disarmament. This is one reason Steve Chapman, in a 2002 *Chicago Tribune* commentary about "flawed" nuclear policy, reminisced that "Bush campaigned on the need for a whole new approach to nuclear strategy [overcoming the rigid dogmas of the Cold War]. Maybe someday he'll provide it."

Many former Reagan-Bush officials resented the fact that the United States, despite its immensely superior nuclear arms, could not control world events (e.g., suppressing "rogue states" and international terrorism). Brandishing nuclear weapons is a risky strategy. With the advent of the G.W. Bush presidency, these holdovers took office and again flexed their policy muscles.

Under Bush, a newfound evangelism was actuated against nuclear proliferation. Nonproliferation became a U.S. policy adjunct intended to help the administration prevail against designated rogue states. Indeed, if Iraq had obtained nuclear weapons, it might very well have been able to stave off an invasion. The situation has different with regard to North Korea; even without nuclear weapons, the DPRK has had formidable conventional forces positioned in range of Seoul, South Korea. In Iran's case, nuclear weapons in their possession would make them even less "manageable."

In essence, the second Bush Presidency national-security trademark seems to have been reasserted by a cabal of hard-liners who persisted from the first Bush Presidency. Public manifestation came in the form of the Project for the New American Century, a group of hard-line conservatives who were associated with the Reagan and Bush administrations and who strongly supported the militarization of Israel (see following boxes). Their platform of across-the-board unilateralism, nuclear bravado, and pre-emption became imprinted in the 2002 National Security Strategy. As G.W. Bush administration policies materialized to be exceeding expensive and unpopular, the Project faded into Internet oblivion.

[Project for the New American Century — Statement of Principles (1997)].

[As] the 20th century draws to a close, the United States stands as the world's preeminent power. Having led the West to victory in the Cold War, America faces an opportunity and a challenge: Does the United States have the vision to build upon the achievements of past decades? Does the United States have the resolve to shape a new century favorable to American principles and interests?

We seem to have forgotten the essential elements of the Reagan Administration's success: a military that is strong and ready to meet both present and future challenges; a foreign policy that boldly and purposefully promotes American principles abroad; and national leadership that accepts the United States' global responsibilities.

Of course, the United States must be prudent in how it exercises its power. But we cannot safely avoid the responsibilities of global leadership or the costs that are associated with its exercise. America has a vital role in maintaining peace and security in Europe, Asia, and the Middle East. If we shirk our responsibilities, we invite challenges to our fundamental interests. The history of the 20th century should have taught us that it is important to shape circumstances before crises emerge, and to meet threats before they become dire. The history of this century should have taught us to embrace the cause of American leadership.

Our aim is to remind Americans of these lessons and to draw their consequences for today. Here are four consequences:

• we need to increase defense spending significantly if we are to carry out our global responsibilities today and modernize our armed forces for the future;

• we need to strengthen our ties to democratic allies and to challenge regimes hostile to our interests and values;

• we need to promote the cause of political and economic freedom abroad;

• we need to accept responsibility for America's unique role in preserving and extending an international order friendly to our security, our prosperity, and our principles.

Such a Reaganite policy of military strength and moral clarity may not be fashionable today. But it is necessary if the United States is to build on the successes of this past century and to ensure our security and our greatness in the next.

Habituated to Cold War Thinking

Seemingly stuck in a time warp (during both the Clinton and G.W. Bush administrations), much of post-millennium U.S. national policy, international strategy, military funding, and intelligence organization mimicked outmoded thinking. An insightful analysis by Russian foreign-affairs analyst Alexei Arbatov, formerly a member of the Duma, expressed fears that more nations will follow the Cold War example of superpower nuclear deterrence.

Testifying in 2004 before the commission investigating the 9/11/2001 terrorist attacks, CIA Director George Tenet admitted:

We all understood bin Laden's intent to strike the homeland but were unable to translate this knowledge into an effective defense of the country.

It will take us another five years [from 2004] to have the kind of clandestine service our country needs.

Despite demise of the Soviet Union as a classic enemy, U.S. intelligence was unable to successfully apply resources, people, or ability to focus on new and more complex threats, some of direct danger to the homeland. Internal coordination was not particularly improved through three presidential administrations. At least four different terrorist-identity databases were set up between the departments of State and Defense, the CIA, and the FBI, but none was "interoperable or broadly accessible."

[The Project for the New American Century] (www.newamericancentury.org).
...is a non-profit educational organization dedicated to a few fundamental propositions: that American leadership is good both for America and for the world; and that such leadership requires military strength, diplomatic energy and commitment to moral principle.

The Project for the New American Century intends, through issue briefs, research papers, advocacy journalism, conferences, and seminars, to explain what American world leadership entails. It will also strive to rally support for a vigorous and principled policy of American international involvement and to stimulate useful public debate on foreign and defense policy and America's role in the world.

– William Kristol, Chairman

A prime example of this dysfunction is found within the text of the declassified presidential daily briefing of 6 August 2001. The box on the following page has much of the information assembled about Osama bin Laden by the intelligence services; it was released to the public because of pressure from the 9/11 investigating commission. Evidently large pieces of the September 11 puzzle were in the hands of disparate and disjointed bureaus of the U.S. government. Yet it confirms that the FBI was indeed sharing information with the CIA, which assembled and composed the president's briefing.

[Signatories of Project for the New American Century Statement].
Elliott Abrams, Gary Bauer, William J. Bennett, Jeb Bush, Dick Cheney, Eliot A. Cohen, Midge Decter, Paula Dobriansky, Steve Forbes, Aaron Friedberg, Francis Fukuyama, Frank Gaffney, Fred C. Ikle, Donald Kagan, Zalmay Khalilzad, I. Lewis Libby, Norman Podhoretz, Dan Quayle, Peter W. Rodman, Stephen P. Rosen, Henry S. Rowen, Donald Rumsfeld, Vin Weber, George Weigel, Paul Wolfowitz

One conclusion to draw from a Clinton/Bush presidential-policy review is that nuclear-deterrence thinking permeated post-Cold-War thinking, with habitual reactions to classic national-security dangers, whereas the arising terrorist threat should have required a new paradigm in policy guidance.

George Kennan's reflections on nuclear deterrence are precisely described in the following statement from his Spring, 1987, article in *Foreign Affairs*:

When I say that this [nuclear-deterrence] military factor is now of prime importance, it is not because I see the Soviet Union as threatening the United States or its allies with armed force. It is entirely clear to me that Soviet leaders do not want a war with us and are not planning to initiate one. In particular, I have never believed that they have seen it as in their interests to overrun Western Europe militarily, or that they would have launched an attack on that region generally even if the so-called nuclear deterrent had not existed.

This analysis was ripe for the Cold War; Kennan would be among the first to warn that it has outlasted its usefulness.

["Bin Laden Determined to Strike in U.S."] Clandestine, foreign government, and media reports indicate bin Laden since 1997 has wanted to conduct terrorist attacks in the US. Bin Laden implied in U.S. television interviews in 1997 and 1998 that his followers would follow the example of World Trade Center bomber Ramzi Yousef and "bring the fighting to America."

After U.S. missile strikes on his base in Afghanistan in 1998, bin Laden told followers he wanted to retaliate in Washington.

An Egyptian Islamic Jihad (EIJ) operative told that bin Laden was planning to exploit the operative's access to the United States to mount a terrorist strike.

The millennium plotting in Canada in 1999 may have been part of bin Laden's first serious attempt to implement a terrorist strike in the United States. Convicted plotter Ahmed Ressam has told the FBI that he conceived the idea to attack Los Angeles International Airport himself, but that bin Laden lieutenant Abu Zubaydah encouraged him and helped facilitate the operation. Ressam also said that in 1998 Abu Zubaydah was planning his own U.S. attack. Ressam says bin Laden was aware of the Los Angeles operation.

Although Bin Laden has not succeeded, his attacks against the U.S. Embassies in Kenya and Tanzania in 1998 demonstrate that he prepares operations years in advance and is not deterred by setbacks. Bin Laden associates surveyed our embassies in Nairobi and Dar es Salaam as early as 1993, and some members of the Nairobi cell planning the bombings were arrested and deported in 1997.

Al Qaeda members — including some who are U.S. citizens — have resided in or traveled to the U.S. for years, and the group apparently maintains a support structure that could aid attacks. Two al-Qaeda members found guilty in the conspiracy to bomb U.S. embassies in East Africa were U.S. citizens, and a senior EIJ member lived in California in the mid-1990s.

A clandestine source said in 1998 that a bin Laden cell in New York was recruiting Muslim-American youth for attacks.

We have not been able to corroborate some of the more sensational threat saying that Bin Laden wanted to hijack a U.S. aircraft to gain the release of "Blind Sheikh" Omar Abdel Rahman and other U.S.-held extremists.

Nevertheless, FBI information since that time indicates patterns of suspicious activity consistent with preparations for hijackings or other types of attacks, including recent surveillance of federal buildings in New York.

The FBI was conducting approximately 70 full-field investigations throughout the United States that it considered bin Laden-related. CIA and the FBI are investigating a call to the U.S. embassy in the UAE saying that a group or bin Laden supporters was in the United States planning attacks with explosives.

****

Although the superpowers manufactured huge arsenals accompanied by cataclysmic strategies for their use, neither the number of weapons nor the implicit mutual danger has significantly diminished despite the new post-Cold-War climate. Strategy doctrinaires were slow to recognize that national security had to be readjusted to new realities. Massive, excessive nuclear arsenals became a burden that inherently sustains unnecessary risk of unauthorized or accidental use. In the meantime, other ways of inflicting mass casualty have become attainable worldwide.

Previously, the main goal of nuclear-weapon states was to solidify their hegemony by brandishing arms and promoting nonproliferation; now greater emphasis is on reducing arms and inhibiting all weapons of mass casualty. While waiting for agreements on verifiable cutbacks, a number of declaratory policies could be implemented to decrease the dangers of hasty retaliation to false, accidental or unauthorized missile launches.

Nuclear policy is not limited to national security. Political interests have overtly become the driving force for former superpowers in justifying retention and even expansion of versatile and operable nuclear arsenals. In fact, domestic politics often trumps international political issues.

National missile defense was heavily promoted by the G.W. Bush administration. To have success against ballistic missiles, several criteria must be met: realistic assessment of threats and responses, a defensive posture integrated into a constructive arms-control strategy, technical and political feasibility, and overall affordability. The larger goal is to minimize genuine threats to national security. In Section G, a negotiated defense initiative is proposed that could better meet such conditions.

Pressure brought about by outsiders continues to be the main strand of continued resistance to ill-conceived, ideologically motivated programs that fund new nuclear-weapon systems, prematurely deploy a Potemkin ballistic-missile defense, and preserve superfluous nuclear armaments. Outsider opposition remained important as the Bush administration, expressing its notions of unilateralism, was slow to learn constructive lessons from the past.

Moreover, the veil of government secrecy, a common bureaucratic concealment mechanism and a frequent government tool to stifle dissent, was uncloaked about justifications put forth for the invasion of Iraq. To vindicate its unprecedented April 2003 invasion of Iraq, the Bush administration had portrayed the threat to the United States as certain, imminent, and frightening — assertions subsequently found to have been mostly bogus.

Bush-administration allegations about WMD and terrorist connections were not substantiated by either classified intelligence information or after-the-fact intrusive inspections of Iraq. Though not disclosed until months after the invasion was over, the House Intelligence Committee found that Congress and the public were duped by the Executive Branch.

Although an oppressive Iraqi dictatorship was toppled, the fractured nation is faring not much better than the fragile government of Afghanistan. Because of conspicuous U.S. domestic political agendas, widespread skepticism has persisted about Bush-administration motives for sustaining the notion that— to compete effectively against terrorism— "wars" are needed, as in previous and unsuccessful hyped wars against drugs, poverty, hunger, and AIDS.

The most typecast *bête noire (black beast)* in *G.W.* Bush administration mythology has been Iran, frequently categorized as a *"rogue"* state. Although ostensibly its nuclear-enrichment program is not (yet) in violation of non-proliferation norms, Iran has not helped matters by stonewalling international inspections. That Iran has made sufficient headway to possibly threaten Israel's Mideast nuclear-weapons monopoly has caused Israel—perhaps in tacit collusion with the Bush administration—to look at pre-emptive options (see box).

Some other factors relevant to the post-Cold-War nuclear-policy debate are outside control of the nuclear-weapon states. Blunt overwhelming military force is not well suited for suppressing terrorists and terrorism, as the Israelis in Palestine and the British in Northern Ireland have found out in nearly everlasting bloody and costly ways. Nor are residual nuclear arsenals of much use in quelling domestic, regional, or international hostilities.

Nationalism, often camouflaged as patriotism, is a political and psychological conception of the past few centuries; nationalistic fervor has been of immense value to demagogues and tyrants. Nationalism—sometimes overriding democratic values—transcends cultural, ethnic, and geographic diversity as sustainable properties. The most fervent of statists still deny that self-determination fever led to the Balkan conflicts both before and after the Cold War. Continued retention of massive nuclear arsenals under the guise of national security seems to reflect a similar form of convoluted thinking.

[Israel Readies Forces for Strike on Nuclear Iran] (Published online late 2005).

Israel's armed forces have been ordered by Ariel Sharon, the prime minister, to be ready by the end of March [2006] for possible strikes on secret uranium enrichment sites in Iran, military sources have revealed.

The order came after Israeli intelligence warned the government that Iran was operating enrichment facilities, believed to be small and concealed in civilian locations....

Iran's stand-off with the International Atomic Energy Agency (IAEA) over nuclear inspections and aggressive rhetoric from ... the Iranian president ... are causing mounting concern.

A senior White House source said the threat of a nuclear Iran was moving to the top of the international agenda....

Defence sources in Israel believe the end of March to be the "point of no return" after which Iran will have the technical expertise to enrich uranium in sufficient quantities to build a nuclear warhead in two to four years.

"Israel — and not only Israel — cannot accept a nuclear Iran," Sharon warned .... "We have the ability to deal with this and we're making all the necessary preparations to be ready for such a situation."

The order to prepare for a possible attack went through the Israeli defence ministry to the chief of staff. Sources inside special forces command confirmed that "G" readiness — the highest stage — for an operation was announced last week.

[The] head of the Atomic Organisation of Iran, warned yesterday that his country would produce nuclear fuel. "There is no doubt that we have to carry out uranium enrichment," he said.

[Although] Iran insists it wants only nuclear energy, Israeli intelligence has concluded it is deceiving the world and [Iran] has no intention of giving up what it believes is its right to develop nuclear weapons....

Cross-border operations and signal intelligence from a base established by the Israelis in northern Iraq are said to have identified a number of Iranian uranium enrichment sites unknown to the IAEA.

Since Israel destroyed the Osirak nuclear reactor in Iraq in 1981, "it has been understood that the lesson is, don't have one site, have 50 sites," a White House source said.

If a military operation is approved, Israel will use air and ground forces against several nuclear targets in the hope of stalling Tehran's nuclear programme for years, according to Israeli military sources.

It is believed Israel would call on its top special forces brigade, Unit 262 — the equivalent of the SAS — and the F-15I strategic 69 Squadron, which can strike Iran and return to Israel without refueling.

"If we opt for the military strike," said a source, "it must be not less than 100% successful. It will resemble the destruction of the Egyptian air force in three hours in June 1967."

Benjamin Netanyahu ... said that ... "when I form the new Israeli government, we'll do what we did in the past against Saddam's reactor, which gave us 20 years of tranquility."

B. TREATIES AND NEGOTIATIONS

Formal written commitments made in the name of the United States have the force of law. Under Article VI of the Constitution, "... all treaties made, or which shall be made, under the authority of the United States, shall be the supreme law of the land...." Consequently, formal pacts get close scrutiny by the Senate. The United States has entered into a wide range of bilateral, multilateral, and international accords, as well as less-formal executive agreements, that relate to weapons of mass destruction.

The Executive Branch negotiates and signs treaties on behalf of the United States. Usually it is the Department of State that negotiates and the President who signs. To have all of its provisions fully implemented, a treaty must be ratified by the Senate.

In Russia, "legally binding agreements are the only mechanisms that actually spur legal, regulatory, and procedural change in the Russian system." Without a binding treaty, Russian law prevents changes in various military procedures, one example being shipboard training and certification of the handling and maintenance of nuclear weapons — even if not deployed.

Supporting the need for written understandings was a brief 2001 controversy alleging Russian redeployment of tactical nuclear weapons to the Kaliningrad Baltic Sea enclave: Although tactical reductions have been carried out voluntarily under bilateral post-Cold War presidential nuclear initiatives, unclarifiable confusion occurred as to alleged presence of Russian nuclear weapons at Kaliningrad. The United States eventually had to drop its vocal complaints because it lacked substantive legal standing and verification rights. (The enclave has been mentioned as a logical part of a proposed nuclear-free-zone in central Europe.)

Although priorities vary from nation to nation, both nuclear and non-nuclear states have vested interests in achieving peaceful accommodation and readjustment. Arms-control and nuclear-disarmament agendas are becoming increasingly proactive in order to extend, enable, and solidify ongoing normalization and projected reductions, especially with increased worldwide concern for proliferation in weapons of mass casualty.

The United Kingdom credits itself, through the late 1990s, with ratification or negotiation of the following: South Pacific Nuclear Free Zone, Comprehensive Nuclear Test Ban, strengthened international safeguards, NPT review process, security assurances to non-nuclear-weapons states, fissile cut-off, South East Asian nuclear-weapon-free zones (NWFZ), Central Asian NWFZ, international response to India and Pakistani nuclear tests, North Korean safeguards obligations, and IAEA inspections in Iraq.

The United States has entered into and abided by a long list of formal agreements. After the Cold War ended, new treaties have entered into force and others are being negotiated.

Treaties can be classified into five groups: nuclear-umbrella alliances, mutual nuclear restraints (on testing and development), multipartite nonproliferation, bilateral basing agreements, and nuclear-related activity (including nuclear-weapon-free zones).

For some treaties, like INF and START I, their formal functions are winding up. Although still applicable under the scope negotiated, their primary tasks of verified mutual arms destruction have been or are about to be finished. As called for under the 1991 START I agreement, the last of 149 U.S. Minuteman silos was destroyed in August of 2001.

Because the G.W. Bush administration wanted to unilaterally develop and deploy a ballistic-missile defense that would have exceeded the provisions of the ABM Treaty, that venerable accord and others were placed in jeopardy. START II, whose entry into force was effectively put on hold, became a casualty of "unilateralism," as did other arms-control agreements, past and future.

Nonproliferation and Arms Control

First, one must take note of the status of treaties or negotiations aimed at limiting further proliferation or controlling the growth of arms (in contrast to reversing proliferation or reducing arms).

NPT. The centerpiece of international control over the spread of nuclear weapons is the 1968 Non-Proliferation Treaty, which was renewed indefinitely in May 1995 when 185 nations signed the extension. As of 2003, India and Pakistan were two newly minted nuclear-weapons-states not signatories to the NPT; Israel has been the sole Mideast state to refrain; and isolated Cuba was the fourth and only other nonsubscriber nation. Every other member-state of the United Nations has become a signatory of the NPT.

Despite the few holdouts, the NPT is the most universal treaty ever put into force — measured in terms of the number of signatories and in terms of the degree of adherence. No nation directly violated the treaty, although Iraq circumvented the intent and specific terms and was a few years away from making nuclear weapons. North Korea formally withdrew from the NPT in 2003 before it confiscated previously safeguarded plutonium fuel.

One long-evident lesson from the experience with Iraq was that the IAEA inspection mandate was indeed too limited; member states did not authorize random inspections of facilities, especially facilities not declared to be part of peaceful atomic-energy enterprises.

Except for the five original nuclear-weapons states, all other members have formally renounced nuclear weapons; in fact, Jonathan Schell observed that they "manage very well without fission bombs, fusion bombs, and [MIRVs] and give no thought to nuclear deterrence, mutual assured destruction, flexible response or any of the rest of it."

Nuclear-weapon-free zones have also been expanded: The original zone in South America was established by the Treaty of Tlatelolco, and an area of the South Pacific was added under the Treaty of Rarotonga in 1986. When Ukraine, Belarus and Kazakhstan signed the NPT in 1992, they immediately renounced nuclear weapons (and their burdens); ten years later, five Central Asian states agreed on a treaty. When South Africa dismantled its nuclear arsenal in 1991, the entire continent became officially nuclear-weapons-free. Mexico has spearheaded an effort to raise the profile for de-nuclearized zones, while Israel has been a drag on headway in the Mideast.

Every five years, starting in 1995, NPT review conferences are held. At the onset of the millennium, the 188 members (having gained three signatories) gathered in New York. While some good-faith progress to eliminate nuclear arsenals has been made by nuclear-weapons states, they continue to be criticized for the slow pace of disarmament. Rarely has the United States accorded enough respect to send the President, Vice-President, or Secretary of State to the conferences.

> **Article VI.** Each of the Parties to the Treaty undertakes to pursue negotiations in good faith on effective measures relating to the cessation of the nuclear arms race at an early date and to nuclear disarmament, and on a Treaty on general and complete disarmament under strict and effective international control.

The most contentious NPT obligation is Article VI, written into the treaty as a major inducement for non-nuclear states to renounce their right to develop nuclear weapons — even while the five nuclear powers retained their own. At the 1995 renewal conference, the five nuclear powers reaffirmed their obligation to move away from weapons for their own security, including "determined pursuit ... to reduce nuclear weapons globally, with the ultimate goal of eliminating those weapons...."

Ten years after the indefinite 1995 NPT extension, little progress exists in actual or negotiated nuclear reductions. In fact, soon afterwards the United States implemented PDD 60, a policy that retained nuclear weapons as the cornerstone of American security for the foreseeable future. Ambiguity in formulation of NPT Article VI had provided the original rationalization for hawks to retain nuclear weapons until *after* general and complete disarmament was to have been achieved.

Absent progress on implementation of Article VI, nations like India have justified their own resumption of nuclear-weapons testing. Although claiming that threats from China (and implicitly Pakistan) were the proximate reason for testing, the failure of major powers to make significant movement toward nuclear disarmament provided an excuse for newcomers to develop their own capability.

> **[United States Scuttles International Non-Proliferation Reforms (2005)].** Despite increasing concerns over illicit nuclear weapon networks and terrorists seeking weapons of mass destruction, negotiators working on a reform package to beef up the United Nations failed to agree on how to revamp global non-proliferation rules.
>
> They adopted a watered-down package of reforms to be endorsed by the leaders of the world attending the 60th anniversary meeting of the global body.
>
> Proposed new rules on nuclear weapons proliferation and disarmament were completely disregarded.
>
> "It's a real disgrace," said UN Secretary-General Kofi Annan, lamenting the omission, which reportedly came after Washington gave only lukewarm support for the reforms.
>
> He blamed "posturing" for the failure to find a common approach to the spread of weapons of mass destruction.
>
> Annan called nuclear non-proliferation and disarmament "our biggest challenge, and our biggest failing," citing a similar failed effort at an earlier NPT conference.
>
> John Bolton, ex-arms control chief at the U.S. State Department and Bush-administration ambassador to the UN, reportedly was against the proposal initially and, some claim, had campaigned against it. Lukewarm U.S. support for disarmament efforts stems from concerns relating to issues such as the Comprehensive Test Ban Treaty, which Washington refused to ratify.

[The Case for U.S. Article VI Compliance]. At the seventh (2005) Nuclear Non-Proliferation Review Conference in New York, the U.S. representative to the conference stated that the United States was adhering to "Nonproliferation Pact" and pointed to its record on reducing nuclear weapons, which put it in full compliance with its nuclear disarmament obligations. The following is an excerpt from the State Department's press release.

Ambassador Jackie Sanders, the special representative of the President for the nonproliferation of nuclear weapons, noted that even though the Cold War that shaped the Non-Proliferation Treaty (NPT) is over, new threats have arisen to international peace and security. Particularly ominous, she said, are the growing proliferation challenges from the "threat of terrorism involving the use of weapons of mass destruction, the alarming examples of certain [governments] violating their solemn nonproliferation commitments by seeking to acquire nuclear weapons or the means for their production, and revelations of non-state actor involvement in the illegal trafficking of sensitive nuclear technology."

The NPT came into effect in 1970, when the United States and Soviet Union were nuclear-armed adversaries whose nuclear weapons programs were in full force. The treaty aimed to provide a collective security framework in which nearly 190 governments would undertake commitments — on a reciprocal basis — to prevent the spread of nuclear weapons beyond the five which already acknowledged having them.

In response to critics who claim that the United States has not done enough to comply with Article VI, Sanders provided a list of U.S. nuclear weapons reductions:

* dismantled more than 13,000 nuclear weapons since 1988;
* reduced deployed strategic warheads from over 10,000 in 1991 to fewer than 6000 by December 2001, under START (Strategic Arms Reduction Treaty) auspices;
* will reduce operationally deployed strategic nuclear warheads to between 1,700 and 2,200 by the end of 2012 — an 80 percent reduction since 1991 — as codified in the 2002 Moscow Treaty on Strategic Offensive Reductions;
* will shrink its overall nuclear stockpile by almost one-half from the 2001 level by the end of 2012 — about one-fourth its size at the end of the Cold War;
* reduced nonstrategic nuclear weapons in NATO by nearly 90 percent since 1991, including reducing types of nuclear weapons from five to one, and nonstrategic weapon storage sites by 80 percent;
* reduced the number of U.S. nonstrategic nuclear weapon systems from 13 to two;
* no longer deploys Navy surface ships with nuclear weapons; and
* dismantled the last of 3000 tactical nuclear warheads in 2003.

Ambassador Sanders added that the U.S. disarmament record also includes notable achievements in reducing nuclear delivery systems, controlling fissile material, cooperating on threat reduction efforts, maintaining a moratorium on nuclear testing, removing bombers from stand-by alert, and ending daily targeting of any nation with nuclear weapons. She also pointed to a reduced U.S. reliance on nuclear weapons as evidenced in the 2001 Nuclear Posture Review.

Weaponized nations have given assurances — with a long list of exclusions — that they would not use nuclear weapons against non-weapon signatories of the NPT, but the exclusions are so convoluted that the assurances are considered inadequate by most nations. The norm of international law, according to non-proliferation expert Josef Goldblat, should be "no-use" of nuclear weapons, as already the case for chemical and biological weapons. Retaliation with nuclear weapons would thus become the exception to the general rule. Goldblat suggested that a multilateral treaty should be created for the non-use of nuclear weapons, a treaty open to all states and effective only upon ratification of all acknowledged nuclear-weapons states. As part of the treaty, it would be necessary to destroy short-range nuclear weapons which are essentially first-use implements because when deployed in the battlefield they become "use-them or lose-them" weapons.

General George L. Butler's view of the NPT was that "ingrained patterns of interaction between the nuclear and non-nuclear weapons states are promoting a coming train wreck, a collision of competing expectations that at this juncture are irreconcilable." A "clear and unequivocal commitment" to the goals of Article VI is "essential to give substance to that long-standing declaratory position [eliminating nuclear weapons]."

Fissile Material Cut-Off. At the turn of the millennium, eight nations were known or believed to have produced fissile materials for nuclear weapons: United States, Russia, Britain, France, China, India, Pakistan, and Israel. North Korea has since acknowledged production of weapon-grade plutonium and uranium. (Other parties to the NPT are already under obligation not to produce such materials.)

A universal halt in deliberate production of weapon-grade fissile materials has been widely considered an important gesture for nonproliferation.

The "fisban" is the working name of the proposed formalized halt in production of fissile material, the acronym referring to a Fissile Materials Cut-off Treaty.

The United States, Russia, France, and the United Kingdom have announced that they no longer produce fissile materials for nuclear weapons, and it is generally thought that China has also ceased such production. Lacking is any formalization of this *de-facto* freeze.

The proposed cut-off would affect production and stockpiles of weapons-grade fissile materials. (However, nuclear-weapon states only put forth for discussion those stockpiles that they considered "surplus" to their needs). Inclusion of all producers was considered to be important, and the addition of new (e.g., India and Pakistan) and emerging nuclear-weapon states has been discussed. The problems of dealing with undeclared nuclear-weapon states (*e.g.*, Israel) made negotiations quite complicated.

A fisban poses several problems and limitations. A fissile cut-off would be effective in limiting the size of nuclear arsenals only if accompanied by the systematic demilitarization of weapon-grade fissile materials (see Section E). The ban would be more convincing if disposal took place under international safeguards. Another thorny issue has been to get declarations of pre-existing stocks; these were shrouded in secrecy.

Verification of fissile non-production (at reactors and separation facilities) would be relatively easy because of telltale symptoms. Verification of demilitarization appears to be achievable, leaving little uncertainty. The IAEA has offered to substantiate adherence.

Verification need not be a show-stopper, despite Bush-administration concerns; it could be resolved in a technical forum separate from the principle treaty negotiations.

Complicating the treaty scope were attempts to include other related materials, i.e., tritium; which is not fissile but essential in fusion-boosted and two-stage thermonuclear weapons. Some academic and NGO interveners wanted to include reactor-grade materials — though they are not known to have been used in nuclear arsenals.

As a consequence of the Soviet breakup, fissile materials that qualify as weapon-grade have been produced in reactors still operating or recently shutdown in the newly-independent states. Although vulnerable to theft, these materials for the most part were not being generated for weapons, and thus they require no formal cut-off in production; however, their fuel discharge, storage, and treatment need to be securely managed and safeguarded.

One of the first transitional steps could be the establishment of a comprehensive international register of HEU and plutonium stocks, both weapon and reactor-grades.

Special difficulties arise in making declarations of past production because such declarations tread on sensitive nuclear information and residual nationalistic attitudes. Perhaps the biggest problem, a consequence of complications in assessing actual inventories and losses, is the unwillingness and inability of the nuclear-weapons states to certify their production records and their existing tally. The reluctant nations do not want to disclose sensitive information about the amount of material used in weapons, and they do not want to admit inability to account for significant "losses" (to piping, tanks, waste, and spillage — rather than theft).

Judging by the rhetoric, some of this reluctance derived from deliberate Bush-administration foot-dragging to avoid committing to termination of fissile production. Earlier resistance came about during the Clinton administration, which had an agenda that confused military and peaceful uses of fissile materials.

Because of intransigence by the United States, and gradual realization that a resurgence in fissile production is unlikely, arms-control priority has inevitably diminished for the fisban. In addition, the G.W. Bush administration did not support CTBT entry-into force and withdrew from the ABM Treaty — steps that run contrary to the insistence of other nations that a fisban be negotiated in parallel with other agreements on nuclear disarmament and outer space.

While continuing to openly champion the fisban, Washington has had "specific elements" of the treaty under review. In order to help move the negotiations ahead, China had dropped its earlier demand, opposed by the United States, for concurrent negotiations regarding weapons in outer space.

CTBT. Widely understood in 1995 as essential to the indefinite NPT extension was the promise to negotiate and put into force a Comprehensive Test Ban Treaty.

United States rejection of the CTBT in 1999 was considered by arms control advocates as "a major setback to the nuclear non-proliferation regime and U.S. credibility"; it was "the first time that [the Senate] has defeated a security-related treaty since the Treaty of Versailles nearly 80 years ago." President Clinton in September 1996 was the first world leader to sign the CTBT. Former editor Mike Moore of *The Bulletin of the Atomic Scientists*, a bell-weather arms-control publication, considers the rejection to have been a wake-up call, because "nuclear weapons are as vital as ever."

Up to then, considerable nuclear-laboratory and military-leader support had been voiced for the universal test ban. The weapons-laboratory directors in their official statements wrote, "While there can never be a *guarantee* that the [nuclear-weapon] stockpile will remain safe and reliable indefinitely without nuclear testing, we have stated that we are confident that a fully supported and sustained stockpile stewardship program will enable us to continue to maintain America's nuclear deterrent without nuclear testing." One reason for their confidence was that the eight existing, tested, and validated U.S. nuclear-weapon designs were sufficient to manufacture thousands of present-day warheads.

The JCS and heads of military services joined in supporting the treaty. A CTBT would have no meaningful effect on functional testing of most nuclear-weapon components — such as high-explosives that initiate an implosion in the primary, missile launching and guidance systems, and the 6000 or so parts in a warhead outside the nuclear package.

Six "safeguards" considered necessary for U.S. ratification of the CTBT had been approved by the Senate:

(1) conducting a "Science-based Stockpile Stewardship" program to ensure a high level of confidence in the safety and reliability of nuclear weapons in the active stockpile,

(2) maintaining modern nuclear laboratory facilities and programs,

(3) sustaining basic capabilities to resume nuclear test activities,

(4) continuing a comprehensive R&D program to improve treaty monitoring capabilities and operations,

(5) retaining development of a broad range of intelligence gathering and analytical capabilities, and

(6) understanding that the President would be prepared to withdraw from the CTBT if necessary.

Despite the permanent nature of the treaty, the latter safeguard enables any party to exit from the treaty if "supreme national interest" were jeopardized.

Michael May, former director of Livermore weapons laboratory, raised five questions that he thought need to be answered about the CTBT:

(1) Is U.S. security maintained?

(2) Is the Non-Proliferation Treaty strengthened?

(3) Will the scientific uncertainty regarding nuclear weapons be resolved?

(4) Will the stockpile stewardship program be maintained?

(5) Can verification be made effective?

May's own provisional answers were affirmative enough for him to offer qualified support to ratification. However, other former heads of the weapons laboratories have privately and reflexively voiced opposition to the treaty.

Regardless of wide support and agreed safeguards, it is evident that the 1999 Senate CTBT-rejection ballot was based on political rather than national-security considerations. Partisanship was strong, partly because in the words of the Arms Control Association "the Republicans so distrust and so despise President Clinton that they're quite willing to inflict damage to Bill Clinton even it means damage to U.S. national security." Also, the Association saw the vote as reflecting opposition to "all arms control," and that the Republicans "actually are anxious to dismantle much of the existing arms control framework as possible." (This indeed seemed to reflect words and actions of vocal members of the G.W. Bush administration.)

Nobel laureates in physics spoke out in support of the CTBT through the American Physical Society. They agreed that "detailed, fully informed technical studies have concluded [that] continued nuclear testing is not required to retain confidence in the safety and reliability of the remaining nuclear weapons in the United States' stockpile, provided science and technology programs necessary for stockpile stewardship are maintained." Moreover, the Society's Council (representing 40,000 academic, industrial, and laboratory physicists) endorsed the CTBT "including its extensive technical and procedural provisions to verify compliance with treaty requirements." A group of 32 Nobelists signed a statement declaring it "imperative" that Congress approve the treaty to "halt the spread of nuclear weapons."

Former chairperson of the Joints Chiefs of Staff, General John Shalikashvili, appointed by President Clinton to review the CTBT Treaty, voiced strong support for the treaty in his January 2001 report. Shalikashvili put forth recommendations to address the concerns of the Senate, including a more integrated non-proliferation policy, enhancement of nuclear-testing monitoring, and strengthened stockpile stewardship. He urged ratification, warning that delay would make it more likely that other nations would "move irrevocably to acquire nuclear weapons or significantly improve their current arsenal."

The co-discoverer of plutonium, Glenn Seaborg, called it a "tragedy" that the United States had failed to rise above the issue of CTBT inspection. Richard L. Garwin, who had contributed to the development of fusion weapons and continued to be a consultant to the nuclear-weapons laboratories, championed the CTBT. He pointed out that

> Even opponents of the CTBT generally support the Non-Proliferation Treaty, extended indefinitely in 1995.... A condition of that extension was the negotiation of a CTBT.... The NPT has played a major role in holding the number of states possessing nuclear weapons to eight instead of the dozens foreseen within a decade in the 1960s.

Inasmuch as all five major nuclear-weapons states have been adhering to the unratified CTBT, provisional application of the treaty — under Article 25 of the 1969 Vienna Convention on the Law of Treaties — enables the essential CTBT monitoring-system components to function. The CTBT cannot fully enter into force until all 44 specified countries sign and ratify.

General Shalikashvili, in his role as Special Advisor to the President (Clinton), deemed that "the CTBT verification system, together with our own national monitoring capabilities [is able to deter cheating by leaving a potential violator with little confidence that they could escape detection]."

The General's confidence in CTBT ratification was reinforced by a National Academy of Sciences report looking into technical issues related to the CTBT. The Academy judged that

> the United States has the technical capabilities to maintain confidence in the safety and reliability of its existing nuclear-weapon stockpile under the CTBT, provided that adequate resources are made available....

Although the G.W. Bush administration did not submit the treaty for ratification, it contributed funds to the International Monitoring System. During his presidential campaign, Barack Obama pledged to "reach out to the Senate to secure the ratification of the CTBT at the earliest practical date and ... then launch a diplomatic effort to bring onboard other states whose ratifications are required for the treaty to enter into force."

ABM Treaty. The 1972 ABM Treaty was once considered a cornerstone of superpower arms control. Joint U.S./RF commitment on the Treaty was reaffirmed at the 1997 Helsinki summit. However, uninterrupted U.S. development with planned deployment of a national missile defense presented a severe policy dilemma: the Russian government did not believe they could "responsibly accept a limited national missile defense deployment in their own interests."

With Russia balking, the United States had to choose between (1) abiding by the treaty without moving ahead on deployment, (2) negotiating changes in the treaty, or (3) abrogating the treaty. The latter action, taken by the G.W. Bush administration in 2001, was unprecedented and detrimental to the overall nuclear-arms-limitation and reduction process.

One of the most vocal proponents of ballistic-missile defense has been Colin Gray, a hard-liner who blamed "cunning offensive-force planners" for defeating previous BMD deployment with "their deadly vu-graphs." While he conceded that Russia's nuclear-armed forces were not a "proximate" U.S. problem, Gray worried about missile threats from "regional powers." He contended, "it is as certain as anything can be ... that a multi-tiered U.S. BMD architecture would defeat militarily any [ballistic and cruise] missile menace from regional powers." Admitting that he had been "strongly persuaded of the case for BMD" for nearly thirty years, Gray dismissed the argument that "today's BMD options are not militarily very impressive." Labeling criticism of the technology as "beside the point" and "attempts to swamp or evade" BMD as "irrelevant," Gray argued against the ABM Treaty regime because it prevented orbital deployments of BMD-dedicated interceptor missiles or of BMD weapons based on "other physical principles." Gray wanted ballistic-missile interceptors of any type permitted in space orbits.

The Duma, meanwhile, stipulated that Russia would not allow START II to enter into force until the U.S. Senate had approved the September 1997 package of agreements signed in New York, including:

> a protocol extending START II implementation period by five years, a memorandum of understanding identifying the successor states to the former Soviet Union under the ABM Treaty, as well as two agreed statements establishing a "demarcation line" between permitted theater missile defense (TMD) systems and restricted ABM systems.

One suggested compromise on the ABM Treaty would have been to center the U.S. missile defense system at Washington, D.C., which would have been compliant with the Treaty (and match Russia's BMD siting at Moscow). However, this was politically unacceptable because it highlighted inability of missile defenses to protect the entire nation and because it renewed visceral public concerns about being a potential target of pre-emptive nuclear attack.

A better suggestion would have been to negotiate with Russia (and other nations) a new agreement: multilateral missile defense coordinated with reductions in existing offensive weapons. Such an agreement would have avoided the ABM Treaty conflict, would make BMD far more feasible if developed in concert with and operated with the cooperation of other parties, and would reduce the primary threat of accidental or unauthorized launch of existing ballistic missiles.

As outlined under the previous subtitle "Arms Control vs. Missile Defense," the Obama administration, not too long after taking office, modified the U.S. ballistic-missile-defense plan and welcomed Russia's cooperation in "a broader defense of our common strategic interests."

Strategic-Reduction Structure

Some treaties not yet in force were intended to reduce strategic arms (see Volume 2).

START II. The START II Treaty fell into limbo, an immediate casualty of G.W. Bush administration's withdrawal from the ABM Treaty. START II was originally signed in 1993 by Presidents G.H. Bush and Boris Yeltsin.

When the Duma ratified START II in 2000, it attached some reservations Russia reserved the right to revoke the treaty if one of several "extraordinary events" took place: U.S. withdrawal from the ABM accord; if nuclear weapons were deployed by NATO on former Warsaw Pact members; and if anything were done to prevent the normal functioning of the Russian early missile-attack warning system.

The Russian articles of ratification recognized that their government would have to provide stable financing of strategic nuclear forces, secure effective national technical means of verification, and ensure the "safe use, storage, elimination and disposal of strategic offensive arms, nuclear warheads and rocket fuel, and to exclude unauthorized access to nuclear warheads."

The treaty would not have been in full effect until the "instruments of ratification" of both the START II treaty and the various protocols signed in 1997 were exchanged between the United States and Russian Federation. In the meantime, the legislatures of both countries seem to have been in a race to see who could do the most to undermine their respective long-range national-security interests.

With the advent of the G.W. Bush presidency the outlook for START II became dim; an alternative was negotiated, the Bush-Putin SORT agreement in 2002 that resulted in smaller effective reductions and minimal. However, SORT did not address other key START II provisions, such as the prohibition against MIRVs.

START III. Negotiations for additional strategic-arms reductions never did get going seriously enough to hyphenate the post-Cold-War era. The Federation of American Scientists had proposed negotiations toward START III, bypassing START II. Limits of 1000 deployed strategic warheads were suggested, with the understanding that START II reductions would continue. By moving ahead with strategic limitations, the FAS saw this as a means of defusing some differences over the ABM treaty.

In any event, according to Dr. Sergei Rogov, Director of the Russian Institute for U.S.A. and Canada Studies, budget limitations forced self-induced reductions in missile capabilities. The Russian defense budget became about one-tenth that of American military expenditures, and Russian maintenance of its nuclear complex about one-tenth of the corresponding $35 billion U.S. allotment. While savings from START II reductions would be significant for both nations if the treaty were to enter into force, the Russian strategic forces would still continue to deteriorate at the present level of resource allocation. This made the situation favorable for mutual reductions to the proposed START III levels of 1000 warheads or less. But to recalcitrant U.S. hardliners, the financial limitations of Russia paved the way for magnifying American numerical dominance in strategic arms.

START III would also have to deal with "reserve" warheads, which represent a "breakout" threat. The United States has about 3000-4000 nuclear warheads on reserve. Confidence and verification measures that could be incorporated in START III include destruction of MIRV buses, administration of data exchanges, and expansion of mutual on-site inspections.

At the Helsinki summit of 1997 between Presidents Clinton and Yeltsin, an understanding was reached that a ceiling of 2000 to 2500 strategic nuclear weapons for each side would be met by the end of 2007, irrespective of START II ratification. The presidents also agreed that START III would include measures relating to the transparency and destruction of nondeployed inventories of strategic nuclear warheads. Meanwhile, the START II deadline for warhead eliminations was extended to 2007, although deactivation of those scheduled for elimination was to take place by the end of 2003, a goal not met.

The G.W. Bush administration was reluctant to totally eliminate strategic platforms or warheads. Although recognizing that the United States and Russia "are no longer adversaries," the President set a sluggish goal of (unilateral) reductions in its strategic nuclear forces "to a total of between 1700 and 2200 operationally deployed strategic warheads over the next ten years," in effect postponing reductions.

This meant that warheads removed from deployment would not necessarily be destroyed. This also meant that only strategic — not tactical or reserve — warheads would be affected. For START II, delivery systems were "attributed" with an agreed capability of warheads: a Russian SS-18 that could be loaded with 10 RVs would count for 10 warheads even if it had just one warhead loaded on it. A nation would thus have incentive to reduce strategic-delivery capability. Shifting from "counting by attribution" to "counting by operational deployment" would leave much more capability for rearmament and less incentive to convert to more stable deterrent forces. Thousands of unaccountable nuclear warheads could be kept in reserve.

Another impediment to irreversible reductions was the Bush-administration precept that it would forgo a written contract so that it could undo the agreement at will. When the administration came into office, it did not want to agree to irreversibility of reductions, nor commit to destruction of warheads removed from deployment.

In addition, the administration wanted to rearm while disarming: It was planning to convert ballistic-missile submarines into platforms for Tomahawk cruise missiles (as many as 154 of them), which can be loaded with either conventional or nuclear warheads. In the absence of formal verification, the type of warhead carried by the submarine could not be confirmed. Furthermore, the United States wanted dual-capable B-1B bombers excluded from the nuclear counting rules of START III; the B-1 could be returned to a nuclear strike mission in less than six months with extra B61 and B83 bombs kept in active reserve. Another deliberate loophole in the administration's proposal was that submarines being overhauled would not be accountable.

START III has been essentially lifeless, and no U.S. initiatives were placed on the table by the G.W. Bush administration for verified reductions in nuclear arsenals.

Disarmament Initiatives

Some steps toward disarmament had taken place was without the formal treaty-negotiation process when the Soviet Union was foundering. Presidents G.H. Bush and Mikhail Gorbachev in 1991-92 took parallel actions to remove most tactical nuclear weapons from deployment. That understanding resulted in

▸ apparent denuclearization of U.S. and Soviet ground troops,

▸ removal of nuclear weapons from their surface ships and attack submarines, and

▸dismantlement of many warheads withdrawn from deployment.

One abiding limitation with this important initiative was the absence of explicit verification.

Another difficulty was that removal from deployment did not equate to destruction: The weapons could have been simply stored, available for rapid redeployment.

In fact, by the end of 2004, the United States and Russia were at odds over progress in elimination of tactical nuclear weapons. Russia announced that it had fulfilled its obligations by dismantling nuclear warheads from ground-based missiles and removing tactical nuclear weapons from surface ships and submarines. All were relocated to the Russian Federation national territory, unlike U.S. tactical nuclear bombs (remaining in Europe). Because the 1992 Bush-Gorbachev eliminations were voluntary, no treaty or inspection provisions existed to verify its status.

A progressive aspect of the 1992 parallel initiatives was that strategic bombers were to be removed from day-to-day alert and their weapons returned to storage areas. The United States also promised to terminate some weapon programs, leaving single-warhead ICBM modernization as its sole remaining strategic-missile project.

President G.W. Bush did express interest in direct initiatives, which — if reciprocated — would expedite the disarmament process; however, the absence of explicit verification will continue to loom as a millstone around the process of moving toward significantly deeper cuts. President Obama seems more forthcoming, but he has a major domestic agenda to accomplish as well.

CWC and Other Treaties

The Chemical Weapons Convention is being implemented. That includes its comprehensive and intrusive challenge inspections. A full-scale international venue has been established to fulfill treaty requirements.

Although destruction of stockpiles has frequently fallen behind schedules — because of lengthy environmental studies, safety precautions, and hazardous-material transportation difficulties — chemical weapons are being progressively deactivated and destroyed. In the United States, public opposition has delayed shipment of chemical munitions for the purpose of consolidating destruction at fewer and more remote sites.

Several other treaties that do not deal directly with nuclear weapons have progressed to written proposal, explicit negotiation, or actual implementation stages. One of three treaties for which the Conference on Disarmament had established a working group is for the prevention of an arms race in outer space (PAROS).

A Biological Weapons Convention review had been advanced for consideration, but the United States and a few other nations objected partially on grounds of verification inadequacy. The original convention contained no verification or enforcement provisions, and the disclosure of huge biological-weapons production and stockpiles has hastened the realization that a better international agreement is needed. The G.W. Bush administration called the proposed BWC protocol "hopelessly defective in three major respects": (1) provisions that enable inspectors to learn about national defense, (2) compromise export-control systems, and (3) risks to proprietary commercial information.

Although short of being a treaty, the Missile Technology Control Regime is a voluntary association of member states based on an agreed code of conduct against ballistic-missile proliferation. Objectives of the MTCR regime are to utilize space for peaceful purposes, and not to allow concealed ballistic-missile programs. To curb ballistic-missile proliferation, a set of confidence-building and transparency measures have been agreed upon, but there has been no legally binding mechanism to prohibit development of missile-delivered weapons. A few nations that subscribe to the MTCR have been implicated in supplying parts and equipment to proliferant states like Iraq.

Arms-Control Goals

In 1981, as the Reagan buildup was just abut to begin, a colleague and I reminded readers that arms control is a way to limit "the potential for international violence." Shared-restraint in armaments is "likely to promote international stability by reducing uncertainty and insecurity." Recognizing that defense against a *strategic* nuclear attack had become "impractical or impossible," the role of an effective arms-control treaty is to achieve a "balanced, unambiguous, and verifiable military standoff." This was just before the nuclear-arms race again got underway in earnest.

For a carefully negotiated strategic treaty, we enumerated stabilizing factors, objectives that remain valid to this day. As shown in Volume 1: *Nuclear Weaponry (An Insider History)*, the Reagan buildup greatly increased uncertainty and insecurity, placing our mutual survival at immediate risk. The Reagan policy amplified and accelerated Soviet alarm, causing them to ratchet up its nuclear forces, resume SS-20 nuclear missile deployment, place SS-23 nuclear missiles in East Germany and Czechoslovakia, and move its nuclear-armed submarines closer to coasts of the United States. Both sides increased their military spending.

Fast forward to the new millennium. In April 2000 Russia took the lead in arms control away from the United States when its Duma ratified START II and CTBT and reaffirmed its commitment to the ABM Treaty. The United States suddenly found itself in the company of China as the only two nuclear-weapon states that had not ratified the CTBT. And the U.S. government received widespread public criticism for its unilateral stand on unconstrained missile defenses.

President G.W. Bush's Under Secretary of State, John Bolton, admitted that "It has become fashionable to characterize my country as 'unilateralist' and against all arms control agreements." Reaffirming the U.S. commitment to multilateral regimes that promote nonproliferation and international security, Bolton disagreed with critics, pointing out arms-control treaties (of previous presidents) that the United States abided by: NPT, CFE, CWC, BWC, LTBT, PNET, TTBT, and a number of voluntary non-proliferation "arrangements." Also, Bolton averred that "the commitment to honor our arms-control agreements" was fundamental to the administration's policy.

Nonetheless, critics countered that the United States has been rather selective in its dedication: resisting adherence to a number of newer treaties like the CTBT, rejecting improved verification of the BWC, dragging its heels about new strategic-weapons negotiations, withdrawing from the ABM treaty, and hedging about nuclear testing.

Arms control and nonproliferation are certainly not the end-all, nor need to be the primary focus of national or foreign policy. Nuclear controls can indeed be logical byproducts of bilateral, regional and international relationships. However, there is substantial unfinished business: huge nuclear, chemical and biological arsenals; unstable nuclear retaliatory systems; a widespread residue of radioactive and hazardous materials and byproducts; and a huge inventory of fissile components that could be used for terrorism.

Statutory controls over these remnants might not have to be as formal and detailed as in the past, but they do need to be clearly defined and enforceable. Agreements need to be spelled out as should be any contract, no matter how uncomfortable President Bush might have been if he had reneged on his campaign promises to oppose arms-control treaties.

The minimalist arms-control treaty (SORT) negotiated by Presidents Bush and Putin in 2002 was far short of the press releases that claimed it would "liquidate the legacy of the Cold War." Instead, the agreement erased START II, which would have made substantial verifiable reductions in nuclear delivery systems and warheads; and the new treaty indefinitely postponed the hoped-for START III, which would have brought levels verifiably down to an even lower stage of mutual risk. Moreover, the Bush-Putin accord avoided the opportunity to reduce nuclear-alert levels and failed to decrease tactical nuclear weapons.

SORT's purview was specifically narrowed down to *deployed* strategic-delivery systems, thus affecting only those that are poised at each other on a day-to-day basis. An upside is that it represents some headway in threat posture; downside is that the agreement does not call for the elimination of any warheads. Assuming the United States and Russia each have about 15,000 nuclear warheads, the SORT forces no changes; however, each side might — feeling the burden of an excessive inventory — voluntarily continue to dismantle some of the warheads, but it would be done without mutual verification.

While the Bush administration deliberately ignored criticism of retention, possibly expansion, of substrategic (tactical) weapons, the topic needs to be discussed as part of an overarching arms-control approach in order to build trust, confidence, and security against proliferation and terrorism.

The Bush retreat from global engagement had implications that go beyond formalized arms control: They affected a nation's ability to deter, defend, and react to international terrorism. Cooperation in sharing intelligence, enforcing domestic and international law, and prosecuting perpetrators for crimes against humanity would significantly reduce the ability of terrorist networks to recruit, train, finance, and coordinate their activities across national borders.

Mortgaging the Future. In preparing this book about the Cold War and nuclear weapons, my colleagues and I (American and Russian physicists) examined the successes and failures of many arms-control accords, bearing in mind that treaties are intended to improve national security for each party, while preserving flexibility to respond to new threats from other directions.

Although a prime beneficiary of formal arms control has been institutionalized verification, more features emerge when the terms of a pact are made explicit.

Verification assures that both parties are carrying out the terms of the contract upon which collective security is so dependent. A written commitment to permit on-site inspection and other measures allows direct observation of the withdrawal and destruction of warheads and their delivery systems. It also provides assurance that decommissioned weapons would not be taken out of the country, whether by theft or with official sanction.

If nothing is binding, then one side can hold the agreement hostage to loosely related problems: For example, nuclear reductions could be put on hold or reversed if either side became dissatisfied with the other's ABM policy.

Yet, advocates of minimalism criticized what they eulogized as the "endless" years of negotiations it took to achieve the nuclear testing and arms control treaties (like the Limited Test Ban, SALT II, Intermediate-range Nuclear Forces, START

I, and START II). Minimalists conveniently ignored the circumstances: Those negotiations were prolonged largely because of foot-dragging by reluctant governments and defense establishments who had to be pushed to the table by a worldwide, grass-roots demand for arms control.

Hardliners on both sides of the iron curtain often contributed to the problem. Contentious issues — of absolute verification (U.S. demand) versus unacceptable intrusiveness (USSR position) — provided opportunity for indefinite delay. When both sides finally decided that a treaty was indeed in their interest, all sticking points vanished overnight.

Positions were reversed during the Bush administration, with Russia seeking formality and the United States forsaking verification on the plaintive grounds that inspections would be too intrusive for U.S. facilities. Existing treaties have already permitted us to witness the destruction of intermediate-range missiles, confirm the absence of testing violations, and monitor progress in elimination of strategic delivery systems. Two more treaties could be implemented promptly: the United States has been balking on the Comprehensive Test Ban Treaty and rejecting improvements in the Biological Weapons Convention. These treaties would bring greater mutual security and provide increased confidence through cooperative verification.

An overriding public threat at hand — despite the spurt of transnational terrorism — stems from the large number of nuclear weapons still in the arsenals of at least nine nations. The United States and Russia together have about 30,000 warheads, of which up to 7000 are operational strategic weapons — the rest being spares, reserves, or tactical weapons. The risk of accidents, unauthorized use, or terrorist acquisition correlates with the number of warheads in existence.

Unless the tradition of formalized reduction and verified destruction of the strategic delivery systems and weapons is renewed, the stimulus will dissipate for eliminating (or even restricting) tactical nuclear warheads, thus perpetuating an ever-present threat — a mortgage — on future stability.

Indeed, under different circumstances, earlier arms-control negotiations were (deliberately) prolonged. With improved accommodation between Russia and the United States, and the mutual desire for smaller arsenals, details could in principle now be settled smoothly and rapidly. Furthermore, there is a strong START treaty foundation of negotiation, verification, and compliance upon which to build.

U.S. government foot-dragging on codification of arms control might have an ulterior motive: "to orchestrate a long-term advantage in nuclear weaponry." That's what officials in Moscow suspected, as the G.W. Bush administration avoided commitments to make reductions irreversible. Aside from the ideological underpinnings of U.S. motivations, the idea that complex understandings could be simply decreed in unwritten form left an unavoidable image of amateurism in international affairs.

Handshake accords, rather than formalized treaties, leave a high degree of insecurity. Avoiding treaties to preserve flexibility is a two-edged sword. It might streamline the process, but it means that military planners will have to hedge against surprises, even to the point of unilaterally reversing reductions already underway.

Ronald Reagan prudently adopted the "trust, but verify" principle. If, for internal and national-security reasons, parties to an agreement carry out arms cuts, their verification would still important — it makes the achievement clear to all. Just as nobody can be sure that the 1991 presidential decision to withdraw all tactical nuclear weapons from Europe was entirely fulfilled, no one will know whether future informal deals are really carried out. Moreover, who could predict whether the next administration in the United States or Russia will want to abide by a predecessor's handshake agreement?

To make an agreement legally binding, it needs to be codified in writing so as to explicitly define the agreed conditions.

After assessing the world situation as of February 2002 that caused them to advance their symbolic clock closer to Doomsday, the advisory board of the *Bulletin of the Atomic Scientists* suggested measures that could increase nuclear stability. The *Bulletin's* assessment concentrated on benefitting from the arms-control and verification experience gained during the East-West confrontational years. Since then, more profound derivatives are associated with what has become a global — as distinct from regional or domestic — heritage of comparative peace and stability. Europe, once the most disputatious arena of hostilities, has become relatively quiescent and interdependent. On other continents once vulnerable to proliferation, nuclear-weapons acquisition has slowed down. Multinational transparency, ushered in part by negotiated treaties and hastened by modern communication and travel, has compressed boundaries. Globally conflictive and threatening activities are, at least now, mostly limited to border-state and intra-state armed struggles.

Accepting a long-term mortgage consisting of negotiated arms control and formalized verification has become an institutionalized and, so far, successful way of minimizing superpower and multinational conflict.

Iraq Invasion Postscript 2:
Multilateralism vs. Unilateralism

In this subsection some policy facets and consequences of the 2003 U.S.-motivated blitzkrieg of Iraq are reviewed. Its impact on regional alliances and on international obligations will take years to unfold. The time and cost of rehabilitation of Iraq is apparently going to be much more than the war planners expected, and the disruption of cooperative traditions between friendly nations has been widely noted.

Humanitarian Intervention. A humanitarian basis for intervention in national affairs already exists under the UN charter, and a substantial literature exists to define or limit intervention.

Key criteria involved in justifying intervention — whether international, regional, or *ad-hoc* — include gravity of human-rights violation, clear and objective evidence, basis in international law, and clear urgency. The use of force is expected to be a last resort and proportionate to the goals. International law in conducting the intervention is to be followed, and there should be a planned transition to post-conflict peace building.

UN/NATO. The United Nations did not sanction the 2003 invasion of Iraq; most UN members favored more time to carry out inspections in order to resolve what appeared to be genuine uncertainties. The invasion and occupation did unambiguously resolve the uncertainties, but not in the way expected: It appears that Saddam Hussein had balked at further encroachment because he anticipated that the Bush administration was intent on invading regardless of the validity of allegations. Hussein, in fact, took several measures in an attempt to forestall the intrusion: He issued misleading orders (disinformation) to his forces regarding the use of chemical weapons, he balked at UN destruction of all of his medium-range missiles, and he successfully made it appear that his armed forces were well-equipped and motivated to resist attack.

NATO was marginalized by the American- and British-dominated "Coalition of the Willing," which consisted of 34 nations. Unable to gain approval or support from NATO or the UN, the invasion went forth anyway; all former Warsaw Block members contributed troops. As of March 2004, the United States had 130,000 troops in Iraq; the United Kingdom, 9000; Italy, 3000; Poland 2460, Ukraine, 1600; Spain, 1300; and Netherlands, 1100. Seventeen nations had less than 1000, and the remainder less than 100.

An egregious political/military error was made in assuming that the Sunni-controlled armed forces would abandon their arms and ammunition when they capitulated; to the contrary, the forces of Hussein evidently prepared a contingency plan, realizing they could not withstand an American invasion. The plan called for hiding small-arms and ammo and organizing the armed resistance that plagued Iraq long after its formal surrender.

The United Nations has long offered a collective international mechanism for problem nations like Iraq. In the absence of post-invasion planning, and despite its advertised aversion to nation-building, the Bush administration had to appeal to the UN and to NATO allies for assistance in quelling, managing, and rebuilding an Iraq freed of Saddam Hussein and the Baathic regime.

International Inspections. While UNMOVIC* teams had undertaken sweeping and unscheduled inspections — steered by U.S. intelligence — of suspected sites for banned weapons and production facilities, essentially no violations of UN resolutions were found. It appears that the UN did not miss any evidence in its pre-invasion inspections.

Hans Blix, who led the original UN inspection effort, advised that several lessons can be drawn from the experience. One was that Iraqi disarmament "would have [had] greater international credibility" if the UNMOVIC team could have validated the invading coalition's (negative) findings. (That point became moot when, to the contrary, the Coalition confirmed the absence of WMD.)

Blix also mentioned that long-term monitoring remains available for Iraq; this oversight measure was not rescinded from UN resolutions that referred to a future WMD-free zone. He added that it would be "a little paradoxical" to simply do away with this opportunity, "especially if we are looking for enhanced verification for the region at some stage," not only for nuclear but more comprehensively for biological and chemical weapons.

He also reminded us that the UN Security Council had personnel within UNSCOM and UNMOVIC which constituted a "unique, elite trained [inspection] force." Blix suggested that maintaining a roster of trained inspectors is "a practical and inexpensive way of holding an inspectorate ready" for qualified assignment to other tasks. It would be particularly valuable standby capability regarding future disputes over missile capabilities, a growing priority for which there is no organized international inspection mechanism.

Anti-Satellite Weapons. Inasmuch as orbiting satellites played a major role in precision strikes against Iraq (and, since then, in Afghanistan), an obvious defense is to pre-emptively attack them in orbit. LEO satellites were used to gather near-real-time information on targets of interest, to input geodesic coordinates for targeting, and to provide terminal guidance of missiles and bombs to these targets.

---

* United Nations Monitoring, Verification and Inspection Commission (see Volume 2)

Consequently a nation anticipating hostility would consider and might develop active and passive defenses to nullify the value of the free-ranging satellites. Iraq was prevented from having longer range rockets to launch ASAT weapons, but other nations (including North Korea, India, Israel, and Pakistan) have the rockets.

One means of defense would be to directly attack low-Earth-orbit satellites (of which two were of primary use by the United States in its bombing of Iraq). Another defensive measure would be to disable the sensors by detonating a nuclear-facilitated electromagnetic-pulse weapon to create disruptive radiation fields in space. In the absence of international arms control, development of EMP weapons could become a multinational free-for-all. Actual testing of EMP weapons would leave a ring of lingering radiation in space.

Unilateralism. Neocon proponents of unilateralism, who gained ascendency in the G.W. Bush administration, created a policy based on hegemonic and absolute military dominance. Unilateralists dislike alliances like NATO, are suspicious of international treaties, and are inclined toward pre-emptive warfare. They have stirred up a furor in Europe and much of the world, especially in nations not dependent on U.S. funding or military aid.

The SORT agreement fits into the arms-management-by-news-management style consistent with unilateralism. For example, by implying that normal retirement of aging nuclear warheads represented deliberate stockpile reduction, the administration publicized the Bush-Putin SORT agreement as a formal arms-reduction treaty. It is not; it's only a gentleman's agreement, reminiscent of the Reagan-era "build-down" whereupon some weapon systems were reduced while others were "modernized." Retirement of the W76 warhead, for example, had already been planned by the United States, partly because it has fewer built-in safety features than others in the arsenal. Nor is it publically known whether retired W62s will be rendered permanently dysfunctional. The SORT agreement simply ratifies reductions in obsolete weapon systems without agreeing to any forced or verified eliminations.

Unilateralists vigorously oppose binding and verifiable treaties such as START. In addition to stymying formal arms treaties, unilateralism's go-it-alone approach avoids international organizations and regional consortia. Although *ad-hoc* coalitions can be organized, they tend to have a transient role compared the more enduring United Nations organization. President Bush's supporters had expressed disdain for the United Nations, and for multilateralism in general, but political, economic, and military realities forced policy reappraisal.

A case in point is the administration's Proliferation Security Initiative. Originally conceived by the President as an intelligence-sharing and interdiction-training project, it has been constrained by lack of international blessing. To avoid formalization, it must rely on the individual capabilities of (a dozen or so) nations to interdict shipments transiting territorial sea, land, and air. Lacking international authorization to carry out operations in extraterritorial waters and airspace, the Initiative is legally stymied by the universal right of "innocent passage." Unless internationally sanctioned and broadened to include a much larger and more significant share of nations, the Proliferation Security Initiative adds little to the standard nonproliferation repertoire, which consists of diplomatic approaches, treaty regimes, legalized inspections, export controls, economic sanctions, and military force.

While counterterrorism is benefitting from multilateral cooperation and coordination, results from the largely unilateralist U.S. approach to Iraq and to proliferation have yet to show a constructive outcome.

****

A Cold War framework of arms-control agreements, mostly codified in national and international law, helped deter dangerous and expensive development of power imbalance. The success of this arms-control framework depended on a number of factors, including creation of verifiable conditions. Good faith, of course, entered into the composite, but the strongest motivator remained: national self-interest to avoid deliberate, accidental, or unauthorized nuclear incidents and attacks.

The inherited international-restraint framework addresses the basic capability to produce nuclear materials and weapons and extends to their testing, delivery, and deployment. Even defenses against ballistic missiles had been an integral component of the arms-control structure, which is another reason that unilateral cancellation of the ABM Treaty risked far more than might be gained in the short run by an ephemeral missile defense.

The benefits of formalized arms control have often been challenged, systematically under the G.W. Bush administration. As before, so-called "strong-defense" proponents continued to rail against treaties — such as the CTBT — with contrived arguments about nonverifiability. The underlying motive of the nullifiers is stated to be national self-interest rather than cooperative security. CTBT ratification has been stymied within the United States — which tested more weapons than any, and had more to lose if others tested. Even though prominent individuals outside of government have repudiated the need for further nuclear testing, subcutaneous political pressure of a detrimental and ecumenical nature dominated.

Reluctance to relinquish nuclear weaponry, especially under pressure brought by the weapon-lab constituency, accounts for systemic resistance to the CTBT. Hope and faith by militants for earth-penetrating miracles and BMD marvels explain backroom efforts to revive nuclear testing.

The weapon-states have overtly ceased production of weapons-grade nuclear materials without formal confirmation; yet a treaty to validate fissile-material production cutoff has been tabled for many years. Sometimes, verification difficulties were given as the reason, but cessation of wholesale production is one of the easiest things to confirm. Past-production and inventory are much more difficult to verify, but they are secondary issues. Problems such as extending the cutoff to additional nations and protecting process secrets are also subordinate to a verified production cutoff.

Proudly proclaiming a freewheeling approach — unilateralism for international affairs — a militant faction had broadly denigrated the importance of formal negotiations, of treaties, and of verification. The faction distrusted, disparaged, and rejected arms accords. When holding office, these ultra-nationalists managed to induce U.S. withdrawal from the ABM Treaty, rejected verification provisions for the Biological Weapons Convention, blocked CTBT ratification, and defaulted on START. This retrogression more than symbolized fundamental misunderstanding of the entire rubric of formalized arms control: It reflected contempt for the world outside the western cattle ranges. Nations have not only grown highly interdependent for their conventional economic and military security but have also benefitted from mutualism to impede nuclear destruction out of the blue.

Bush-administration propositions for strategic-arms reductions seemed more like a shell game: The warheads were to be shuttled between operational systems and storage/reserve. Reductions were limited to "operationally deployed strategic nuclear warheads" — yet those not "operationally deployed" can be retained for prompt redeployment in several ways: cruise missiles loaded on submarines, dual-purpose B-1B bombers, ballistic-missile subs being overhauled, or Minuteman-III missiles reverted to multiple warheads. All of this was to happen, under the Bush plan, without further negotiation or verification.

United States repudiation of the negotiation paradigm, and shyness toward international cooperation, came at a time when Russia, more than ever, has great interest in reducing its equally huge arsenal of strategic and tactical nuclear weapons. It was also a time when the existing arsenals were still facing each other. Joint reductions would greatly and immediately improve each nation's respective national security, as well as conform to international norms for progress in nonproliferation and disarmament.

Abrogation of the ABM treaty elated unilateralists, but it portended an increasingly difficult transition into a progressively interdependent world. Columnist Charles Krauthammer trumpeted the withdrawal as though it would remove "absurd restrictions on ABM technology." Yet a confounding problem remains: There is a long way to go before the technology would be viable. A better pathway would be to negotiate a cooperative arrangement with Russia and other interested parties for joint development of missile defenses.

Unilateralists like Krauthammer, abhorring cooperation, prematurely proclaimed the eradication of terrorism because of the original prompt, decisive, and convincing American success in Afghanistan — and they have celebrated the administration's futile exertions in arms control because it avoided involuntary nuclear cuts.

The apocalypse-weary of the world are grateful for any agreements between the United States and Russia, informal or formal, that reduce nuclear arms and improve defenses against nuclear war.

The hallmark of a valid contract is a clear understanding of obligations and the ability to verify whether commitments are being met. President Bush is quoted as saying, that "We don't need arms-control negotiations to reduce our weaponry in a significant way." But we do need a covenant to ensure that explicit commitments are fulfilled. Even business deals between friends are carefully worded and signed, to avoid misunderstandings.

Formalized arms-control benefits should not be dumped in order to support current political expediency, at the expense of continued nuclear uncertainty. A signed treaty with Russia, perhaps "START III" with arms-control procedures spelled out, would provide a stable basis for constructive cooperation in missile defense, for dealing with terrorism, and for solving other critical geopolitical problems.

C. STUFFING THE GENIE BACK IN THE BOTTLE

Even if humankind manages to prohibit nuclear weapons, no one can be sure that they will not clandestinely reemerge: from either limited stocks secretly retained by those now possessing them, or new weapons manufactured under cover by what are now non-nuclear states.

Making restraint even more difficult is the fact that nuclear weapons are symbols of power. The regional struggles between India and Pakistan have been punctuated with tit-for-tat nuclear tests, accompanied by continuing border disputes. North Korea's antediluvian leadership has chosen nuclear weapons to gain an edge in regional relationships and stature.

Official U.S. nuclear-arsenal policies, to "reduce and hedge" and to be able to fight a "protracted" nuclear war, recall much Cold War thinking — so much that concerns about nuclear-weapons proliferation and accidental or unauthorized use have been "distinctly secondary, according to Wolfgang Panofsky." The highly influential and respected Panofsky warned that, ironically, "the threat of nuclear proliferation is largest to the United States among all other nations [with] the most to lose if nuclear weapons proliferated."

Panofsky also pointed out that potential proliferants can deliver small numbers of nuclear weapons in many ways: "nuclear weapons could be detonated on ships [that come into] harbor, delivered by light aircraft [that fly below radar screens], smuggled [in vehicles] across U.S. boundaries, as well as by ballistic and cruise missiles...." He noted that "meaningful defense against such a spectrum of delivery options is impossible...."

These drawbacks have not discouraged certain ideological factions from proselytizing nuclear weaponry as the pillar of national security. Achieving a "nuclear rollback" would require a fundamental decision by political leadership, realizing that national security is itself threatened by nuclear weapons in the hands of too many parties and that other nuclear-weapons states will reduce their dependence only if the United States cuts its arsenal.

Unfortunately, the NPT and its associated agreements were not designed to deal with a multinational nuclear rollback; new verification measures would be needed to protect the interests of nations that begin the reduction process. In order for the superpowers to reach the lower bounds of reductions, similar assured progress would have to be made by other weapons and threshold states. Assuming the IAEA would have an important function in verifying nuclear rollback, the agency would need openness and transparency among relevant nations.

Another factor detrimental to massive reductions is the lingering potential for weaponization of dormant non-nuclear nations. This is best illustrated by the long nuclearization road traveled by North Korea, and by the amount of hedging carried out by Iran.

Incidentally, Sweden once had an extensive nuclear explosive-program — reversed on its own volition — thus being an exemplar of the genie being put back in the bottle. The Swedish experience is a reminder that advanced industrial states could make nuclear weapons, although it would probably take several years to consummate the endeavor. Had the Manhattan Project not come up with nuclear weapons, they would very likely have been invented later; the concepts and raw materials were previously discovered, and their potential had already been recognized.

For the superpowers to engage in significant nuclear reductions, other nations, those with and those without nuclear weapons, must be jointly involved; this point is discussed in detail in the following four Sections. First, let's scope out the nuclear capability of other weapons and non-weapons states.

Acknowledged Weapon States

To consider possible rollbacks, we have to assess two categories of acknowledged nuclear-weapons states: original members and latecomers.

U.S. and Russia. Leadership of these two nations is key to a drawdown in arsenals. Much of this book is devoted to their role. Reductions in nuclear arms by the superpowers would be partly dependent on reductions and accommodations made by other weapon-states. Also to be factored in are various security umbrellas to which these superpowers are committed.

China, France, Great Britain. All three of these nations maintain relatively small, diversified retaliatory nuclear forces at levels that some would say should be emulated by the two superpowers. China probably retains a retaliatory capability as long as it senses a nuclear threat from Russia, India, or the United States — or a future concern about Japan.

As long as Russia depends so heavily on its nuclear arms, France and Britain, having already reduced their arsenals, consider a nuclear component essential to stability in Europe. Regarding NATO, besides less-ambitious plans for rapid membership expansion, some slackening of nuclear emphasis has taken place in its declaratory policy and by limiting foreign weapon storage in member nations.

After decades of reservations, France has gradually embraced a series of arms-control measures: subscribing to the NPT (1991), halting production of weapons materials (1992-1996), ending nuclear testing, signing the CTBT (1993-1996), adopting protocols for nuclear-weapon-free zones in South America and Africa, reducing its nuclear stockpile (1996), and de-targeting its nuclear weapons (1997).

For multilateral drawdowns to take place in strategic arms, stability in conventional forces and alliances in Europe and Asia would be essential. While Germany has continued to refrain from developing nuclear arms, it has been highly dependent on security umbrellas provided by NATO and particularly the United States.

In order to encourage deep cuts by the superpowers, China, France and Britain would have to refrain from expanding their arsenals or testing nuclear weapons — favorable conditions that are nearly *de facto* now. Through multilateral or international negotiations, such reassurances could be put in writing. In addition, multinational agreements on softer issues, such as de-alerting, de-mating, and no-first-use would help allay fears and reluctance to move ahead with nuclear reductions.

India/Pakistan. Wars and skirmishes have taken place between India and Pakistan ever since they were decolonized by the British, who left artificial borders that failed to reconcile religious and ethnic divisions. These conflicts sparked both nations to seek nuclear weapons. India was able to obtain support for nuclear capability from the Soviet Union, and Pakistan received nuclear and missile technology from China.

The United States turned a blind eye to the transfer because "in the shadowboxing of the Cold War era ... Asia was vital to American interests." In 1971, when India was about to overrun western Pakistan, the United States under President Nixon intervened — in part by dispatching a nuclear-powered (and presumably nuclear-armed) aircraft carrier to the Gulf of Bengal.

While both subcontinent neighbors express policies of deterrence, their deep-rooted antagonism suggests that nuclear weapons could very well be used in an all-out war between them.

Border disputes with China linger as present-day India feels obliged to maintain a nuclear deterrent. Cross-border difficulties with both Pakistan and China provided a rationale for India to come out of the closet as a declared nuclear-weapons state, despite its vocal advocacy of universal nuclear disarmament.

Having now openly declared their nuclear status, the question has come up about how the two new powers would be accommodated by the international community. Neither had subscribed to the NPT nor invited IAEA safeguards. One suggestion has been that India and Pakistan could announce that they will behave as though they were weapon states within the NPT and not transfer nuclear weapons to any recipient nor assist any nation in acquiring them. France took the same official position before it signed the NPT in 1992. India and Pakistan could also reconfirm their moratorium on nuclear-weapons testing and accede to the CTBT once (or if) the five major weapons states have ratified it.

For India and Pakistan to cultivate mutual confidence and roll back their nuclear armaments, they would probably want to adopt intrusive regional and international verification arrangements that would oversee their production facilities, stored weapons, and specialized delivery systems. Before this could be accomplished, significant slackening of tension over internal, border, and regional disputes and terrorism would have to occur.

North Korea. In attempting to keep the Democratic People's Republic of Korea within the NPT regime, the United States in 1994 negotiated an agreement providing fuel oil, with a promise of building two light-water power reactors to provide energy self-reliance. If, in exchange for political and economic incentives from other nations, North Korea would have agreed to intrusive safeguards of their entire nuclear program, there would have been little risk of that nation embarking on nuclear-weapons development or long-range ballistic-missile improvement

Internal political survival has been crucial to the North Korean leadership while the world about changed has much more rapidly. The unsettled truce between North and South Korea is another factor creating instability.

Until President G.W. Bush labeled it as part of an "axis of evil," it appeared that North Korea could be bought off. But, in apparent reaction to Bush-administration interventionist policies, the North Korea government acknowledged a program of nuclear-weapons development, including the removal of Yongbyon reactor fuel rods previously under IAEA safeguards. Subsequently it publicized progress in processing the fuel into weapons form. Ultimately North Korea carried out two series of underground nuclear explosions.

Considered to have been one of the destination nations for centrifuge enrichment cascades illicitly marketed by Pakistan, North Korea has since made public its uranium fissile-separation capabilities for weapons. Clearly regional security and economic issues would need to be resolved before the DPRK would backtrack on its nuclear capacity.

Threshold Weapon States

In this category are included nations that have hovered near the nuclear-weaponization threshold. Some briefly crossed over and subsequently retreated, others walked up to, but never quite crossed it; and at least one seems determined to make the full transition. Deliberate ambiguity has been the hallmark of these programs.

South Africa. South Africa's nuclear-weapons program is an appropriate case study in reverting from production to termination. It is the first and only nation certified to have abolished a nuclear arsenal: South Africa indeed stuffed the genie back.

One of their chief program officers, Waldo Stumpf, provided important details about this unparalleled retroversion from an unacknowledged nuclear-weapons state to a disarmed member of the NPT.

It was not until 1993 that South Africa publically admitted the irregularities: The government had clandestinely developed a "limited nuclear deterrent capability" in the 1970s and 1980s. Six of seven planned fission weapons were built and destroyed before South Africa joined the NPT in 1991, at which time the IAEA was "granted full access to the facilities which had been used in the past [for the surreptitious nuclear-weapons program]."

Termination officially began in 1990, with *sub-rosa* dismantlement of a high-enrichment (90% uranium) fissile-production cascade. Safe and secure disassembly of the nuclear bombs was carried out, and an independent South African expert was appointed to independently audit the process to ensure no diversion of nuclear materials. According to Stumpf, all seven devices were dismantled; the HEU was melted and recast; facility decontamination was carried out; factories were converted to conventional-weapons work; and all design drawings, manufacturing information, and hardware were destroyed. Dismantlement was essentially complete by June 1991, allowing good-faith accession to the NPT.

Upon invitation, the IAEA later sent a verification team. After extensive inspections and analysis, the IAEA verified the HEU enrichment plant's inactive status. The General Conference of the IAEA "accepted the completeness of South Africa's inventory of materials and facilities ... and declarations on the dismantlement and destruction of the hardware from the [seven] nuclear devices."

After the democratic election of Nelson Mandela in 1994, the government "committed itself to a policy of transparency with regard to the nonproliferation of weapons of mass destruction." It is now a domestic criminal offense "for any South African citizen to develop, or assist in the development of, chemical, biological and nuclear weapons as well as ballistic missiles capable of delivering such weapons." South Africa is seeking declaration of the entire continent as a nuclear-weapon-free zone.

Israel. For motivation to eliminate its nuclear force or potential, Israel would have to be very satisfied with security conditions in the Middle East. Even if other nuclear nations eliminated their weapons, Israel is unlikely to be influenced enough to forego the nuclear option without its national identity assured.

Were peace and security to come to the Middle East and Israel, the assumed rollback of its unacknowledged nuclear arms could probably be conducted in the same fashion as South Africa, that is, through an internal process, ultimately verified by the IAEA. Stocks of fissile materials would eventually have to be transferred, converted, or consumed.

Iraq/Iran. Evaluating Iraq's compliance after the first Gulf War, Rolf Ekeus reported that "Iraq has no nuclear weapons, and with the existing monitoring regime, Iraq will not be able to acquire nuclear weapons." UNSCOM inspectors had found a nuclear military program, but weapons were many years away from being fabricated. Iraq's case was unique in that nuclear-materials production capability was physically destroyed by international intervention.

[Relevance of Israel's Osirak Destruction to Events Regarding Iran (2006)]. A recent analysis of Israeli destruction of the Iraqi Osirak reactor in 1981 agrees that the raid "almost certainly" speeded up the Iraqi nuclear program. "The raid was a tactical success but a strategic failure." From a course of nuclear-weapons development that might have taken ten or more years through plutonium diversion, the Iraqis embarked on a "much larger and ambitious" uranium-separation program with an order of magnitude increase in personnel. Israel apparently remained "unaware of Iraq's new efforts throughout most of the 1980s."

This history is relevant to current events regarding Iran. Some neoconservative analysts propose keeping the nuclear-facility attack option open, believing they need "the threat of military action to force Iran into compromise [on its nuclear program]."

As a result of the subsequent 2003 invasion, it is likely that outside scrutiny and prohibitions will be imposed for a long period of time.

In Iraq's case, pre-Gulf-War motivations for acquiring nuclear weapons had more to do with regional issues, namely security or aggression; Saddam Hussein probably expected that his possession of nuclear weapons would prevent intervention in his attempted takeover of Kuwait and his other planned military ventures.

If Iran is engaging in a similar nuclear-militarization program, it might have analogous justifications. Although its existing program ostensibly is directed at peaceful uses of nuclear power, the Iranian government has hedged by moving uranium enrichment close to the threshold of weaponization when not under intense international safeguards. As of this writing, only compliance with IAEA access demands would meet non-proliferation concerns (see status reports that follow).

Of course, one of the major secular factors now affecting Iran has been its need for deterrence to restrain Israel. (Previously it had more to do with its hostile relationship with Iraq.) Conversely, Israel has been concerned that its implied nuclear deterrence is insufficient.

The complexities of sub-national terrorism, mixed in with Mideast insecurities, compounded by religious and factional differences, make it difficult to orchestrate multilateral accords.

[Iran's Nuclear Program (2006)]. Although Iran has been party to the NPT and the IAEA inspection additional protocols, suspicions have evolved over its nuclear intentions.

Iran has complied with the NPT and the IAEA enhanced-inspection regime, while the United States and other nations have failed to fulfill their nuclear-demilitarization commitments under Article VI of the treaty.

Iran had implemented inspection protocols that allowed access to facilities and suspended its uranium-enrichment program, pending resolution. The IAEA inspection team for several years reported no evidence of prohibited programs or materials, although Iran has since reverted to non-disclosure.

Worldwide there is considerable misunderstanding with regard to weaponization of low grades of fissile materials. While methods for production or isolation of these materials can, after renouncing international safeguards and treaties, be extended to produce ample stocks of weapons-grade materials, declared or undeclared nuclear weapon states have taken that route because of the immediate visibility and expected reaction.

Iran has sought peaceful and pollution-free nuclear energy for its rapidly growing nation; it wants to conserve oil resources in order to obtain hard-currency revenue. These are increasingly sensible policies that were originally encouraged by the United States. Under the NPT, Iran has the inalienable right to develop peaceful applications of nuclear energy.

The Iranian government has admitted that it resorted to concealment and blackmarket purchases. It blamed the West because of imposed restrictions on its nuclear program, which Iran insists was for legitimate peaceful application of nuclear power.

Moreover, the government of Iran is aware that possession of nuclear weapons could undermine its security. Iran can rationally justify peaceful coexistence with its neighbors, especially if the Mideast were made into a nuclear-weapons-free zone.

To reinforce nonmilitary intentions, Iran had offered to submit its entire nuclear-fuel cycle to outside scrutiny, to restrict uranium enrichment well below weapons quality, and to undertake technical, regulatory and legislative measures. In return for its transparency, cooperation, and voluntary suspension of its program, Iran expected concessions in accordance with a November 2004 agreement in Paris reached with the European Union.

In 2006, because of an impasse in negotiations with the European Union, the government of Iran withdrew from its agreement for additional protocols. It continued to be a party to the NPT and allowed normal IAEA safeguards inspections of declared locations.

Iran's resistance to foreign interference opened the door for the G.W. Bush administration to steamroll European Union and United Nations sanctions, to incite the resistance of Iranian mullahs, to manufacture incidents and negative publicity, to orchestrate pre-emptive action by Israel (see next box), and to justify potential American nuclear brinkmanship.

[In 2009 Iran acknowledged existence of an underground enrichment facility, without supplying details to remove *prima-facie* ambiguities regarding its purpose and capabilities. It might be intended (and would be suitable) for enrichment sufficient to supply its American-built medical and scientific research reactor in Tehran.]

[Lessons from Iraq Can Help Deal with Iran (2009)]. Where are the diplomatic carrots for Iran? Ever since the 11 September 2001 terrorist attack, U.S. national security has been used as rationalization for many domestic and foreign transgressions, not the least of which was invasion of Iraq based on fictitious allegations about weapons of mass destruction, export of terrorism and ties to Al Qaeda.

During the G.W. Bush administration, we heard of U.S. government preparations for using nuclear weapons against Iran.

The evolving scenario was chillingly familiar: We could expect the president to solemnly address the nation and say that it is necessary to use nuclear weapons in the name of national security against a threat (to American interests) perceived to come from Iran. This announcement would likely follow some (easily contrived) incident.

Important lessons from invading Iraq were not being taken into consideration: to keep an open mind, to maintain channels for negotiation and to avoid policies that label nations as evil.

Here are some relevant policy factors:

■ Iran has complied with both the Non-Proliferation Treaty and the International Atomic Energy Agency enhanced-inspection regime, while the United States and other nations had failed to fulfill their own nuclear-demilitarization commitments under Article VI of that same treaty.

■ Iran had agreed to and implemented inspection protocols that allowed access to its facilities and, pending resolution, had once suspended its uranium-enrichment program. The IAEA inspection team for several years reported no evidence of prohibited programs or materials, although Iran has now reverted to non-disclosure.

■ Iran is no more likely than Iraq to tolerate outside intervention. Faulty, incomplete or misunderstood intelligence data should not be a basis for nuclear policy.

■ Iran has long sought peaceful and pollution-free nuclear energy for its rapidly growing nation, seeking to conserve its oil resources in order to obtain hard-currency revenue. These are increasingly sensible policies that were originally encouraged by the United States. Under the Non-Proliferation Treaty, Iran has the inalienable right to develop peaceful applications of nuclear energy. Iran is being held to a standard that nuclear-weapon states brush themselves aside.

■ Iran can rationally justify peaceful coexistence with its neighbors, especially if the Mideast were made into a nuclear-weapons-free zone, a policy that the U.S. could promote. The government of Iran is well aware that possession of nuclear weapons could undermine its own security.

■ Iran had offered to submit its entire nuclear-fuel cycle to outside scrutiny and to restrict uranium enrichment so that it is only useful for peaceful nuclear power (fuel enriched well below that needed for weapons). Iran had also offered to undertake technical, regulatory and legislative measures to register its nonmilitary intentions. If Iran is disrupted from legitimate pursuits or its hand is forced, the United States is no more likely to succeed in forestalling nuclear weapons than it had been with North Korea.

Argentina/Brazil. Both of these bordering South American nations have voluntarily receded from the weaponization threshold. They have signed a nuclear accord, established the ABACC for verification by mutual inspections, are parties to the Treaty of Tlatelolco, and have become signatories of the NPT. Thus, whatever might have once existed as a nascent program of weaponization has already been voluntarily reversed.

Non-Proliferation Policies

Ever since the genesis of fission explosives, weaponized states have attempted to constrain others' proliferation. Unfortunately, what is beneficial for one nation's security is not necessarily suitable to another. Policies that depend heavily on the production and threat of using nuclear weapons hardly beget restraint from others.

Proliferation has been motivated more by domestic politics and external pressures than technical capabilities. Support for this deduction comes directly by analyzing the motivations of NPT signatories, which have voluntarily adhered to its terms, while the few non-signatories/exemptions have deliberately obscured or exercised their nuclear options. In any event, nuclear policies that contravene the NPT are patently ineffective in promoting nonproliferation.

Mislead environmentalists have been on a wild-goose chase. At one time, with nuclear weapons derived from the same fissile source materials, and with a couple of economically catastrophic accidents, there was ample reason to question the future of peaceful nuclear power. But now, with nuclear power's substantial and lengthy record of safe, secure, and economic energy, peaceful applications have endured and proven themselves through more than a half century — a remarkable and telling record in the annals of human industry.

The earnest concern of environmentalists would now be better aimed at existing stockpiles of *weapon-grade* fissile materials and nuclear arsenals. As previously pointed out, most power reactors offer essentially no prospect of aiding weapons proliferation because of inherent design features and internationally applied safeguards. The few that could make separable weapons-grade fissile materials are closely safeguarded. Although the NPT bans military uses, it encourages peaceful applications of the atom, a partitioning that should be respected in the interests of international stability. Russia and other nuclear-capable states export peaceful nuclear technology, particularly to complete or follow-up on nuclear-power reactors exported by the FSU.

So what should be a sensible non-proliferation policy toward a wannabe? Jonathan Schell has articulated an appropriate proliferation-nonproliferation conundrum. As a result of North Korea's movement toward nuclear weapons, the G.W. Bush administration "discovered that its policy of pre-emptively using overwhelming force had no application against a proliferator with a serious military capability, much less a nuclear power." The lesson that North Korea and Iraq received from Bush-administration policies is "if you really want to defend yourself, develop nuclear weapons, because then you get negotiations, and not military action."

Schell tracked pre-emptive war policy as far back as a statement from General Groves, overseer of the Manhattan Project, whom he quotes: "If [any foreign power with which we are not firmly allied] started to make atomic weapons, we would destroy its capacity to make them before it has progressed far enough to threaten us."

Israel twice attempted that against Iraq, only to channel and accelerate Iraq's efforts along a more promising clandestine proliferation route.

Schell inferred that "nuclear nations acquire nuclear arsenals above all because they fear the nuclear arsenals of others." He pointed to the incentive of "Hitler's phantom arsenal" inspiring "the real American one." And, he observed that "the Soviet Union built the bomb because the United States already had it." As for England and France, they reacted to the Soviet threat; China responded "to the threat from all of the above," Pakistan answered India, and North Korea has reacted against the United States (and its surrogate, South Korea).

Schell attributed President Bush's revival of preemption to his all-out "war on terror" after 9/11, even though so-called "realists" (including iconoclastic John Mearsheimer) were prepared to live with Iraq nuclear-armed, but contained.

Potential proliferators see not only "the hypocrisy of great powers delivering sermons on the virtues of nuclear disarmament while siting atop mounts of nuclear arms," but also the beleaguered governments visualize an increasingly "dangerous neighborhood" of adjacent nations. Israel and South Africa — and other threshold nations — saw themselves in an environment where nuclear weapons might be needed as a last (perhaps suicidal) resort.

In order to deal with the proliferating possession weapons of mass destruction, Schell saw a stark choice: either "universal permission" or "universal prohibition." The price to release humanity from the peril of weapons of mass destruction "is to relinquish our own." Schell agreed with another proliferation analyst, George Perkovich, who contended that

> the grandest illusion of the nuclear age is that a handful of states possessing nuclear weapons can secure themselves and the world indefinitely against the dangers of nuclear proliferation *without* placing a higher priority on simultaneously striving to eliminate their own nuclear weapons.

Iraq Invasion Postscript 3:
Non-Proliferation Implications

With the 2003 invasion of Iraq, the prospects of nonproliferation suffered a devastating blow. Nations (such as North Korea) that fear aggrandizement by others were impelled to put nuclear-weapon development on the fast track. That happened before to India, Pakistan, Israel, and South Africa. The invasion of Iraq led by two nuclear-weapon states will undoubtedly cause other nations to undertake or give new consideration to nuclear weaponization.

Take for instance Syria and Iran, who border the zone invaded and occupied by the "coalition of the willing." Surely they will be under considerable internal pressure to acquire weapons widely considered as the ultimate defense. The entire non-proliferation paradigm has significantly changed as a result of the invasion.

Moreover, thinly veiled U.S. threats to *use* nuclear weapons — coupled to a standing policy of *first use*, and lack of any serious move to reduce arsenals — will further devalue nonproliferation. Severely undermined are the fragile international institutions that had evolved to forestall weapons of indiscriminate casualty.

Even more troublesome has been the undermining of internationalism, (1) by dissolving existing treaties, (2) by abstaining from new accords, and (3) by invading Iraq without UN sanction, or approval of the international community. These U.S. policies have run counter to nonproliferation, which depends on substantial international consensus.

****

For superpower nuclear disarmament to make headway, comparable reductions would have to be carried out by other nuclear-weapons states, as well as those nations that have touched the threshold. Historical experience is support of that potential: Some weapon-states have demonstrated self-restraint, rollbacks have been put in place, and (secret) reversal of weaponization has happened. This affirmatory experience shows that nuclear reductions, even disarmament, can take place successfully and later be verified.

As to the more provocative question, whether or not the genie can be stuffed back in the bottle, the answer is **no** *if you are thinking of every bit of nuclear knowledge, technological capacity, and source materials: The physics of thermodynamic entropy teaches us that it is essentially impossible to reverse such processes and conditions to get back to the original ordered state of affairs. But the answer is* **yes** *if you could be satisfied with a more modest, practical goal: reversal of the threat inherent in existence of huge war-fighting nuclear arsenals that are subject to human error, arbitrary judgment, or unmanageable accident.*

D. DEEP CUTS

The threat of nuclear war, which has been an "inescapable backdrop" for about half a century, remains to haunt the world. Newly constituted Russia, inheriting many thousands of nuclear weapons, has been mired in deep economic distress. Strategic nuclear-weapons systems remain on high alert, launchable by accident or mistake. The prospect of additional proliferation lingers. These prolonged hazards have led some observers to "counsel the complete abolition of nuclear weapons" and others to advocate less-traumatic "deep cuts" in nuclear inventories. Less vocal and less dogmatic now are the proponents of massive retaliation.

During the Cold War, negotiated or voluntary nuclear reductions resulted in some numerical (vertical) disarmament, although qualitative improvements in lethality often offset the quantitative calculus. In any event, strategic deterrence envisioned a quantitative floor beneath which it was considered too risky to reduce the tallies. Since then, proposals for "minimum deterrence" have resurfaced, where some smaller number of nuclear weapons — between a few dozen and a few hundred —

would suffice as long as the arsenal could, if necessary, be quickly placed on full defensive alert.

In this Section are compiled and discussed many of the proposals aimed at deep cuts, even abolition, the latter sometimes seen as a "fanciful dream" based largely on moral principles. At the end of the Section, a different (more pragmatic) view is offered. Because of the multiplicity of proposals and factors, they are extracted and collected into separate lists at the end of this Chapter.

Has the Time Arrived?

Making a renewed plea for deep cuts, Jonathan Schell has rhetorically set the stage:

> Who wants these reciprocal threats of annihilation, and why?
>
> Communism was the issue between the Soviet Union and the United States. The government now in power in Russia overthrew Communism, and today relations between Russia and the United States are cordial. The great nuclear arsenals of the United States and the Soviet Union were created as instruments of the Cold War.
>
> Gone is the murderous, implacable hostility between global rivals, which just a few years ago seemed destined to last forever; gone the totalitarian empire; and gone the obstacles to inspection that have been considered the main brake on nuclear disarmament.

Yet, laments Schell,

> threats of "mutual assured destruction" ... remain the order of the day.... During the Cold War, we so accustomed ourselves to threatening nuclear annihilation that it became second nature to us.

The UN General Assembly, whose vast majority of members are non-nuclear-weapons states, found that "the threat and use of nuclear weapons would generally be contrary to the rules of international law applicable in armed conflict, and particularly the principles and rules of humanitarian law" — except, as Schell notes, possibly in "an extreme circumstance of self-defense, in which the very survival of the state would be at stake."

General Lee Butler, former U.S. chief of strategic forces, reminds us of nuclear sufficiency: "Twenty weapons would destroy the 12 largest Russian cities with a total population of 25 million." As an interim compromise en-route to deep cuts, he recommends an arsenal "in the hundreds," but his goal is zero.

The Arms Control Association has placed some strategic issues into the following perspective:

> Whatever value one places on a defense against a highly unlikely emergence of a credible rogue state ICBM threat or on U.S. military actions against states such as Iraq and Serbia, that value pales in comparison with the importance of Russian ratification of START II and the commitment to follow-on negotiations on START III. These treaties would reduce — by about 75 percent from present levels — the number of deployed Russian strategic warheads. In addition they would eliminate all Russian land-based MIRVed missiles, which are particularly dangerous because their high value as targets encourages a precarious launch-on-warning posture. However unlikely a U.S.-Russian conflict may appear today, Russia remains the only country that can threaten U.S. survival.

The eminent scientist Wolfgang Panofsky, a veteran of the Manhattan Project, stated his view about reductions and eventual abolition:

[U.S. national security] for the foreseeable future would be well served by a reciprocal reduction regime leading to a level of "a few hundred" nuclear weapons.... The world cannot afford to postpone action severely restricting the number and use of nuclear weapons until a nuclear detonation actually occurs.

Posed by Panofsky is a rhetorical question:

Can conditions ever be achieved in which the possession and use of nuclear weapons can be prohibited worldwide? I advisedly use the word "prohibited" rather than "eliminated" or "abolished." Nuclear weapons cannot be "un-invented." Thus the best hope for mankind is to arrive at an international norm under which nuclear weapons are prohibited....

During the eight years of the Clinton administration, during which it became clearer that Russia did not pose the same ideological or military threat as the Soviet Union, many proposals emerged for deep cuts. But it soon became evident that dependency on nuclear weapons was a difficult habit to break, and the debate shifted to how deep was "deep." Clinton's Secretary of Defense, William Cohen, a Republican, objected to cuts deeper than 3000 strategic weapons for several reasons:

The Joint Chiefs of Staff during the Clinton administration testified that they would require a major review of nuclear war plans before contemplating reductions below an arsenal of 2000 strategic warheads. The administration's war blueprint was incorporated into the SIOP, which is supposed to contain detailed plans for destruction of 2260 vital Russian targets, some of which required the assignment of more than one nuclear warhead.

Committee on Nuclear Policy

Composed of about three-dozen NGO-project directors, an *ad-hoc* Committee on Nuclear Policy at the onset of the millennium analyzed dangers that existed in Russia from insecure nuclear materials. The project directors and their organizations were all veterans of years of experience with arms-control and non-proliferation issues, many of them having personally spent time in Russia evaluating the situation.

Based on their combined appraisal, the Committee generated recommendations regarding nuclear reductions, transparency, warheads, fissile materials, launch-on-warning, massive-attack war plans, and consolidation of Russia's nuclear materials. Because of the importance of these various recommendations, many of which reflect similar suggestions in this book, they are consolidated them below, starting with concerns about protection of nuclear materials.

Short of endorsing all of their recommendations, they do provide a detailed foundation for serious consideration.

In order to address the most serious of the perceived nuclear-materials weakness, the Committee suggested that funding should be provided to remediate dangers in Russia, citing a number of anecdotes: a guard killed at a nuclear-weapon test site, rampage by a sailor on a nuclear attack submarine, shooting of soldiers at a plutonium storage site, an unguarded facility containing 100 kg of weapons uranium, unpaid nuclear workers, attempted theft of nuclear materials, underfed and underclothed guards at nuclear facilities, shutdown of nuclear-facility security systems, and instability of old ICBMs in hundreds of silos.

The NGO committee was especially concerned about accidental nuclear war risk heightened by hair-trigger alerts. This led them to suggest a stand-down of nuclear forces slated for destruction under START I, joint elimination of launch-on-warning, verifiable removal of nuclear forces from hair-trigger alert, and verifiable elimination of massive-attack options.

The Committee also recommended that existing arms control treaties be supplemented with parallel, reciprocal, and verifiable reductions, with declarations in favor of reductions in deployed strategic weapons within a decade, with cradle-to-grave transparency as the basis for reciprocal reductions, with eventual reduction on total (strategic and tactical) nuclear weapons for each side, and for engagement of other nuclear-weapons states in the reduction process.

Because "effective management of the new U.S.-Russian nuclear relationship also involves differences over the issue of ballistic-missile defenses," the Committee on Nuclear Policy stressed that BMD deployment should be clearly defined, rigorously tested, affordable, cost-effective, balanced, and kept within the context of reducing nuclear hazards.

Staged Reductions

A "technical blueprint for very deep nuclear reductions," has been put forth by coauthors of *The Nuclear Turning Point*, a book contributed by NGO-affiliated individuals who do not believe that "abolition is currently a realistic goal."

The first, most immediate step they encourage is to reduce nuclear risks from accidental launch or detonation, malicious deeds, or impetuous reaction. Operational downgrading alleviates these risks. Moreover, the timeline for return to a war footing would be stretched out, thus allowing opportunities for deliberation and diplomacy in case of a political crisis.

Much of the remainder of *The Nuclear Turning Point* is devoted to staged reductions in warheads and delivery systems. The book emphasizes that, while other countries are striving to reduce proliferation jeopardy by extending the NPT and signing the CTBT, the United States and Russia continue to hold each other hostage to nuclear assault, "against all common sense." Each deploys thousands of nuclear weapons on high alert, many ready to quickly launch on warning of an incoming attack. The two former adversaries reserve the right to use nuclear weapons in repelling a conventional attack. Among military planners the notion still abides that nuclear weapons can be "realistically used in a wide variety of contingencies." Such concepts are "ingredients for accident, miscalculation, and nuclear escalation" and the risks are "heightened by political instability in Russia and conditions verging on economic collapse in much of its nuclear weapon complex."

The proposed framework leading to deep cuts is in three stages (summarized below), beginning with the "de-alerting" and "de-activating" of weapons. The goal is to dispel fears of large-scale surprise nuclear attack, decimate the danger of accidental and unauthorized use of nuclear weapons, and drastically reduce the possibility of nuclear weapons being used in regional conflicts. To reach these goals, they propose (1) the eventual scrapping of almost all of the world's stockpiles of warheads and fissile materials and (2) the forging of institutional arrangements intended to forestall renewed buildups of nuclear arsenals or proliferation.

As a matter of perspective, keep in mind that the United States alone has about 10,500 active warheads and 12,000 spare nuclear pits from dismantled weapons. Three quarters of the active warheads are assigned to strategic deployment.

The first stage of proposed reductions to 2000 warheads apiece would be combined with de-alerting, and de-activation, thus removing weapons from a quick-response mode, making most weapons unusable for weeks or months. (Note: Measures that entail physical — rather than procedural or electronic — steps are usually more durable and verifiable.)

The boosters of this recommended three-stage reduction process consider these to be the "longest steps" in the direction of abolition "that can be realistically projected under current international conditions." Each stage, elaborated below, is predicated on a mood of stability in bilateral and international relations. A favorable atmosphere would permit reductions to take place; conversely, each stage alone would contribute to stable security relationships.

The First Stage: Reductions to 2000 Warheads. Decreasing deployed strategic warheads to 2000 is central to the first stage, and it is consistent with the outline of START III proposed at the Helsinki Summit. Detailed tradeoffs are suggested to enable each side to maintain overall ceilings.

The Nuclear Turning Point highlighted its first reduction stage with continued adherence to the ABM Treaty (subsequently mooted).

Elimination of tactical nuclear munitions is necessary for meaningful removal from service. The Helsinki statement committed the two nations to exploring possible measures for tactical-warhead elimination to be accompanied by appropriate confidence-building and transparency.

Establishing a comprehensive warhead-verification system is another major part of the first strategic-reduction stage. Looking ahead to reductions, getting the verification system in place as early as possible is very important. This could start by declaring numbers and locations of nuclear warheads, followed by identification of each warhead with a unique serial number.

If the first recommendations were consummated, this would define the key "turning point," the demarcation where the superpowers are positioned and ready to proceed to significant reductions.

The Second Stage: Reduction to 1000 Warheads. Reaching a verified ceiling of 1000 strategic warheads — stored, as well as deployed, in the arsenals of the United States and Russia — is a major goal of the second stage. Bilateral verification and monitoring would be put in place not only for the residual arsenal but also for warhead dismantlement and fissile-material disposition.

Their choice of 1000 warheads for this ceiling corresponds roughly to the current combined inventory of the other nuclear-weapons states. With decreases by the two major nuclear powers, it is hoped that the remaining weapon states will see it to be in their own interest to enter into multilateral negotiations for further curtailments.

Several ways of reaching the 1000 level have been proposed, including the tacit replacement of one leg of the strategic triad with strong conventional forces. Another way is to only retain single-warhead nuclear missiles, which would reduce vulnerability to treaty breakout, as would reductions in numbers of long-range delivery systems, such as cruise missiles and bombers.

The Third Stage: Reduction to 200 Warheads. To reach the lowest goal considered realistic in The Nuclear Turning Point, a residual 200 warheads would be allocated to the most survivable legs of the triad (maybe even a dyad). Most of the warheads would be deactivated in such a way that they could not be reinstalled without sounding an affirmable alarm. Other nuclear nations would be asked to reduce their arsenals so that their combined inventory is comparable to that of the United States and Russia.

Of course, the number 200 is alterable; it is a goal that could be set twice as high or half as much. But the authors of *The Nuclear Turning Point* did not choose the number arbitrarily: One factor was their estimate that residual uncertainty in U.S. and RF fissile-material stockpiles would be equivalent to about 200 warheads; another factor was to provide a hedge against breakout by any other nuclear-weapons state.

A Russian's View. Alexei Arbatov — a Russian official who has been closely connected with arms-control activities — added his own pragmatic, staged approach by addressing several "core" issues: deployed strategic forces, undeployed warheads, and tactical nuclear forces.

Arbatov's initial scenario would conform to the "natural decommissioning of technically obsolete weapons." (This is a useful expression of a valuable concept.)

In discussing reserve warheads, Arbatov identifies issues that need to be faced. U.S. reductions under START II were to be implemented by downloading missiles and bombers, while Russia would have to eliminate launchers and missiles. As a result, the United States would have a larger residual inventory of warheads — a breakout advantage. Unless the warheads are destroyed, U.S. missiles could be reloaded, whereas Russian missiles could not accept increased loading.

An alternative suggested by Arbatov favors attrition to eliminate ALCMs and allow verified destruction of RV heat shields. By maintaining a large reserve of warheads "for overt reconstitution of the [nuclear] forces if supreme national interests are endangered," some de-mating agreements might be able to move ahead. The warhead reserve would have to be placed in a large "reconstitution infrastructure" that would be relatively insensitive to pre-emptive strike.

Russian tactical nuclear munitions apparently have design lifetimes that, without servicing or upgrades, would have expired nominally by 2003. An agreement to limit production of new weapons to replace those that are physically obsolete would be propitious to both nations. Deep reductions in such weapons would require verification measures on their elimination and on the disposition of surplus weapon-grade uranium.

The Role of Nuclear-Weapons States

The United States — the nation that first produced, tested, and used nuclear weapons — has special responsibility and self-interest in finding a solution to the nuclear dilemma. One reason is that it has about half the world's nuclear arsenal; another is that America has thought about and agonized over the bomb for more than half a century, being the only nation to invoke their ferocity. Perhaps even more important is the prominence of U.S. cities as potential victims of nuclear destruction. Despite disappearance of the menacing confrontation and the current lack of an extraordinary threat, the superpowers are left with huge dreaded arsenals. In any event, the United States cannot escape its interdependence with Russia and other nations in decisions about the fate of nuclear weapons, and the financial load is harmful to sagging economies.

Britain and France have indicated that the appropriate time for them to join disarmament negotiations is after the United States and Russia have reduced their arsenals to a level of about 1000 weapons each. The other *de-facto* nuclear-weapons states should also be invited to participate in the multilateral reduction process.

If the last of the reduction stages recommended in *The Nuclear Turning Point* were reached, no nation would have more than 200 verified nuclear warheads, and the combined total of Britain, France, and China would not exceed 200. All except the permitted warheads and delivery systems would have been destroyed in a verifiable manner; the remaining authorized warheads would be separated from their delivery systems and subject to international or multilateral monitoring at their dispersed (and defended) territorial storage sites.

The allowed warheads at every reduction stage would be deliverable by means that each weapons state chooses; the most survivable deployment modes for each party are preferred. Warheads would be stored at multiple sites within each nation, often not too far (nor too close) to delivery systems. Storage would be remotely and continuously monitored by the treaty parties, a process that has already been demonstrated for nuclear fuels and materials. Withdrawal of warheads from storage would have to meet agreed conditions.

The objective of monitoring warhead storage is to ensure that — if one nation made a unilateral decision to reactivate the weapons — sufficient warning and reaction time would be available to all parties. Present-day weapon states would probably be able, *in extremis*, to reassemble a militarily-qualified nuclear force within a year or so. Non-nuclear states require substantially more time to create a weapon.

Abolition

Many individuals and NGOs have been promoting the abolition of nuclear weapons. The 1997 National Academy of Sciences CISAC report, which discussed "prohibition" of nuclear weapons, remarked that "The end of the Cold War has created conditions that open the possibility for serious consideration of proposals to prohibit the possession of nuclear weapons." At the time, CISAC could not see clearly how or when this could be achieved; they deliberately chose the word "prohibit" rather than "eliminate" or "abolish" because the members presumed that

the world can never truly be free from the potential reappearance of nuclear weapons [and] the knowledge of how to build nuclear weapons cannot be erased from the human mind.... Even if every nuclear warhead were destroyed, the current nuclear-weapons states, and a growing number of other technologically advanced states, would be able to build nuclear weapons within a few months or few years of a national decision to do so.

Various NGOs, including the Atlantic Council and Pugwash, have been focusing on deep cuts. An umbrella organization, the Abolition Caucus, has urged negotiations toward that end. Individuals, such as retired generals Lee Butler and Andrew Goodpaster, have added their names to petitions. Abolition 2000 formed a network to coordinate organizations appealing for elimination of nuclear weapons, and it published a book, *Security and Survival*, which made the case for a negotiated worldwide agreement on nuclear-weapons abolition.

As quoted from Jonathan Schell, just a few pages ago, "The great nuclear arsenals of the United States and the Soviet Union were created as instruments of the Cold War." That led him to the logical question, "Now that the conflict has been dissolved, can't the arsenals be dissolved?" Tracing the history of abolition attempts, Schell went back to the beginning:

Once the Soviet Union acquired the bomb, in 1949, proposals for nuclear disarmament were rejected on grounds that the character of the Soviet regime posed an insuperable obstacle. Nuclear disarmament, the Cold War catechism ran, was possible only if the arrangements could be fully inspected; the Soviet Union, being a closed society, would not permit inspection of its military establishment; therefore, nuclear disarmament was impossible.

Schell added further justification to universal nuclear disarmament with this logic: "Vertical disarmament makes a catastrophe, should it ever occur, smaller. Horizontal disarmament makes a catastrophe of any size less likely to occur.... Vertical and horizontal disarmament are not mutually exclusive."

Moving to the logical extreme—abolition—is recognition of risks worldwide from human error and misguided intention.

While "abolition" entails the actual elimination of nuclear weapons, "renunciation" is the policy equivalent, a governmental commitment that lacks durability and tangibility. Renunciation would be a necessary but not sufficient step towards nuclear abolition.

As a result of Schell's extensive and detailed interviews, he concluded that

it is unquestionably possible, through technical means, to turn something that is a nuclear weapon into a collection of materials that plainly is not. It therefore would be perfectly accurate to say that when every nuclear weapon has been dismantled to a certain extent, abolition has occurred. And this word, it seems to me, remains fitting whether or not a nation retains the intention to reconstitute nuclear weapons in the event that another nation builds them.

Even though commonplace nuclear threats are reduced as arsenal regression takes place, some nuclear dangers would remain. The crucial difference, Schell reminds us, is that "rather than one nation deterring another, all would jointly deter all from starting back down the path toward nuclear abyss."

Abolition will have to be accompanied not only by political accommodations, but also strong verification, including on-site challenge inspections and reliable means of national intelligence. Moreover, as far as Mikhail Gorbachev is concerned, "We may never be able to solve the nuclear [abolition] question unless at the same time we develop a system of international organizations [including] an effective UN, an effective Security Council and systems of regional security...."

A more recent and surprising convert was Paul Nitze, a very conservative former arms-control negotiator; he began to argue that the United States has such an enormous advantage in conventional weapons that complete elimination of nuclear arsenals could be contemplated.

Military Officers Agree With Abolition

As noted by Jonathan Schell, not only civilians have been alarmed about the slow pace in post-Cold War reductions:

> In the years since the end of the Cold War, a striking number of the men (there were almost no women among them) who had responsibility for nuclear policy in the United States, Russia and Europe have undergone changes of heart about nuclear weapons. They now support a position unthinkable for most them even a decade or so ago — the global elimination of nuclear weapons — and many have come to question the Cold War strategies they once devised.

In 1996 a group of eminent retired military officers from many nations, including the United States and Russia, banded together to call for deep cuts in nuclear weapons to be followed by phased elimination. General Lee Butler was quoted in the *Chicago Tribune* on December 5 of that year:

> As an advisor to the president on the employment of nuclear weapons, I have anguished over the imponderable complexities, the profound moral dilemmas, and the mind-numbing compression of decision-making under the threat of nuclear attack.
>
> I see ... the burden of building and maintaining nuclear arsenals: the increasingly tangled web of policy and strategy as the number of weapons and delivery systems multiply; the staggering costs; the pressure of advancing technology; the grotesquely destructive war plans; the daily operational risks, and the constant prospect of a crisis that would hold the fate of entire societies at risk.
>
> Most importantly, I could see for the first time the prospect of restoring a world free of the apocalyptic threat of nuclear weapons. Over time, that simmering hope gave way to a judgment which has now become a deeply held conviction: that a world free of the threat of nuclear weapons is necessarily a world devoid of nuclear weapons.

Regarding concerns which compelled the preceding judgment, General Butler pointed out:

> First, [Those who think nuclear weapons are usable] don't fully grasp the monstrous effects of these weapons ... poisoning the earth and deforming its inhabitants.
>
> Second, a deepening dismay at the prolongation of Cold War policies and practices....
>
> Third, [deterrence] reigns with its embedded assumption of hostility and associated preference for forces on high states of alert.
>
> Fourth, an acute unease over renewed assertions of the utility of nuclear weapons, especially as regards response to chemical or biological attack.
>
> Fifth, grave doubt that the present highly discriminatory regime of nuclear and non-nuclear states can long endure absent a credible commitment by the nuclear powers to eliminate their arsenals.

And finally, the horrific prospect of a world seething with enmities, armed to the teeth with nuclear weapons, and hostage to maniacal leaders strongly disposed toward their use.

The case for their elimination is a thousand-fold stronger and more urgent than for deadly chemicals and viruses already widely declared immoral, illegitimate ... and prohibited from any future production.

... the real issue here is not the past — it is willingness to undertake the journey.

In General Butler's considered view, three crucial conditions for moving ahead with elimination are:

First and foremost, is for the declared nuclear weapon states to accept that the Cold War is in fact over....

Second, for the undeclared states to embrace the harsh lessons of the Cold War: that nuclear weapons are inherently dangerous, hugely expensive and militarily inefficient....

Third, ... a sweeping review of ... nuclear policies and strategies [is essential].

When asked about some specific arguments sometimes brought against abolition — that getting rid of nuclear weapons will unleash conventional wars, that the weapons cannot be disinvented, that breakout will occur — General Butler responds: "So what? We have had that circumstance already, with nuclear weapons present in the world [mentioning Iraq]." He favors, if necessary, "immediate and unconditional [conventional-force] intervention" to quell regional conflicts.

On conventional wars, Butler remarks, "In truth, in the world of nuclear weapons we have seen some of the most murderous wars in history: Iraq and Iran — more than 100,000 casualties; the Korean War; the war in Vietnam. What did nuclear weapons do to prevent, contain, constrain those conflicts?"

For furthering the control of nuclear weapons, General Butler favors setting up norms of behavior: "systems of enforcement, agreements specifying collective action, capacities for intervention."

Since 1996, General Butler has continued to sustain his strong role in promoting nuclear abolition. *The Bulletin of the Atomic Scientists* has credited him with

the rarest of human traits: the ability to change one's mind about truly fundamental matters as objective circumstances warrant. During the Cold War, he believed the Soviet Union was a threat and he acted accordingly, eventually becoming commander of all U.S. strategic forces. After the Cold War, he recognized a new reality and again he acted. He championed the view that nuclear weapons themselves had become the threat and must be eliminated.

The former strategic commander has more recently underscored and refined his appraisal. He recognizes as "challenging and thoughtful" the point of view often expressed to him by former military colleagues that nuclear weapons are considered the reason that World War III did not break out, and the weapons are believed to stand between the United States and the forces of "barbarism, terrorism, and rogue nations." His shorthand counter-response is, "We in the United States cannot at once hold sacred the mystery of life while we retain the capacity to utterly destroy it."

General Butler is bothered by the "creeping re-rationalization of nuclear weapons — as symbolized ... by the Senate rejection in [1999] of the Comprehensive Test Ban Treaty." Because the "ultimate abolition" of nuclear weapons involves "a long and difficult process as it is," he believes it will become even more difficult the longer the process is delayed.

Gaffney's Response to General Butler. Frank J. Gaffney Jr., formerly responsible for nuclear-weapons policy in the Reagan Defense Department, has responded that nuclear force remains essential to American security.

> An idea that is neither realistic nor, arguably, desirable under present (and foreseeable) circumstances can still be a lousy one, even if some 60 former U.S. and foreign flag officers embrace it.

> ... the retired officers [referring to General Butler and others] are, to put it charitably, imprecise about how the conditions required to make the elimination of nuclear weapons feasible will be achieved. For example, how will the proverbial nuclear "genie" be put back in its bottle? In fact, it is no more practical to believe that the knowledge and means needed to make crude nuclear weaponry can be eliminated worldwide via international treaties and arrangements than it to think that electricity or the internal combustion engine can be disinvented. [Anyone] determined to acquire [nuclear weapons], and able to pay the going rate, can get them.

> ... potential adversaries [are more likely] to try trumping U.S. forces by using weapons of mass destruction— chemical or biological weapons, if not nuclear ones.

> The United States should take not steps that could reduce the disincentives it presents to others contemplating aggression against Americans or their allies and interests.

Other Objections. Even though a great many retired generals and admirals joined General Butler and General Andrew J. Goodpaster in calling for commitment to eliminate nuclear weapons, two prominent retired flag officers did not: Generals Colin Powell and Norman Schwarzkopf.

Objections to deep cuts can also be found in some periodicals: For example, columnist V.H. Krulak, reflecting right-wing opinion, responded that the abolitionist stand of retired military officers confused and complicated the issue, although Krulak agreed that nuclear arms are dangerous and costly. He says that the officers "take no note of the reality that getting rid of all nuclear weapons, even were that possible, would still not destroy the [production] process itself. Nations around the world who were so inclined could still employ the existing technology to create nuclear weapons at will."

To Krulak a threefold "course" for the United States was "clear" because he believes other nations would cheat, not destroying their arsenals simultaneously:

> First, we should forgo any idea that nations of the world will get rid of all nuclear weapons.

> Second, in our own national interest, we should reduce our nuclear-weapons inventory to whatever level we regard as militarily and economically sound, and we should make and keep those weapons in first-class condition.

> Third, we should pursue a technology that will give us a reasonable defense against nuclear attack.

Nothing that Krulak suggested is inconsistent with an effort to gradually abolish nuclear weapons while supporting a sound national defense at all times.

A reasoned objection to deep cuts was put forward by Kathleen Bailey of Livermore weapons lab: "[Because] the value of one or a few nuclear weapons is greatly enhanced when potential adversaries have none ... there would be no way for cheaters to choose not to declare its existing or special nuclear materials and hide them." She compares nuclear weapons to handguns: "You can ban them, but they won't go away."

In extending her gun analogy to nuclear-arms control, Bailey advanced the argument: "if guns are unlawfully used, the state can, if it can find the guilty party, exact punishment. If nuclear weapons are used, however, it is unlikely that an errant nation would be meaningfully punished." (While she has a point, the same problem prevails not just for deep cuts but any level of nuclear armament. The difference is that deep cuts would reduce the possible scale of nuclear exchange that could occur in retaliation — as well as reduce its likelihood of occurrence, while still retaining deterrence.)

According to William Arkin, the lack of "universal support" for General Butler's core idea that "a world free of the threat of nuclear weapons is by necessity a world devoid of nukes," will likely to be found "among groups as superficially diverse as arms controllers, government policymakers, think-tank pundits, and editorial writers." Arkin thinks the naysayers are trapped in an "old circular conundrum" that he succinctly describes as follows:

> What are the targets of nuclear weapons? Nuclear weapons.
> What provocations could bring about the use of nuclear weapons? Nuclear weapons.
> What is the defense against nuclear weapons? Nuclear weapons.
> How do we prevent the use of nuclear weapons? By threatening to use nuclear weapons.
> Why can't nuclear weapons be abolished? Nuclear weapons.

Government Actions and Statements

Various government leaders responsible for nuclear decisions have expressed their views, either during or after leaving office. President Kennedy had well-known reservations about using nuclear weapons.

President Reagan, before ending his presidency, acknowledged that "nuclear war was unwinnable and must never be fought." Both he and Soviet President Gorbachev have publically asserted that "they shared the goal of eliminating all strategic nuclear arms within ten years."

As the Soviet Union dissolved, President G.H. Bush took a calculated political risk by initiating a series of unilateral nuclear-disarmament measures, and President Gorbachev matched Bush's cuts within 10 days.

In 1998, President Clinton gave his view:

> We have an opportunity to leave behind the darkest moments of the 20th century and embrace the most brilliant possibilities of the 21st. To do it, we must walk away from nuclear weapons, not toward them.

Although statements from Chinese officials are often discounted in the West, in 1999 the president of China, Jiang Zemin, wrote that

> For 50 years, hanging over our heads like a sword of Damocles, nuclear weapons have never ceased threatening humanity's survival. The end of the Cold War has not brought about their disappearance.
>
> ... nuclear nonproliferation and nuclear disarmament remain important tasks for the international community.

China's president called for progress in six areas of arms control, as summarized in the next list.

Although the surreal contest between titans is over, in order to avoid a nuclear arms race of potentially even greater magnitude, it behooves the West to pay attention to the suggestions of the much more populated Peoples Republic of China.

As for President G.W. Bush, he kept promises made to his constituency to retain nuclear counterforce as a strategic doctrine and to act unilaterally in world affairs.

Nuclear Obsolescence

Without rigorous maintenance, nuclear weapons and their delivery systems would gradually deteriorate. For example, most modern nuclear weapons, especially strategic weapons, require that their tritium reservoirs be refilled periodically; if not, the tritium will have naturally decayed within 12 years to half of its original strength.

If the capability to manufacture or purchase were scaled back, tritium supply would gradually reach an equilibrium level commensurate with its restocking. Tritium salvaged from dismantled nuclear warheads can replenish reservoirs of the active stockpile.

Other natural processes of deterioration, figuratively called "rust-out," take place in warheads and missiles.

Strategic obsolescence is an issue only if offensive or defensive capabilities were not occasionally rebalanced or upgraded: For example, if one side decided to deploy a proven, effective missile-defense system, the other side might take it to mean a loss in offensive capabilities unless it too took countermeasures or developed alternatives.

Obsolescence has become a rallying cry in itself, an argument to stimulate development and production of weapons and delivery systems. Conversely, by taking the aging process into explicit consideration, gradual erosion in arsenals can occur over a long interval, thus avoiding significant losses of incumbent political capital.

Stockpile-maintenance programs, encumbered to some extent by nuclear-test moratoria, are intended to keep existing weapons in shape to meet strategic objectives. Conversely, such programs can be used as a subterfuge for designing weapons to meet new military missions. That's why objections were raised to Bush-administration programs that would result in bunker-busters and mini-nukes. For example, the program for W76 warhead refurbishment has been tainted by mixed signals and hints about goals other than lifetime extension or improved safety and reliability.

Obstacles to Nuclear Disarmament

Although U.S. presidential administrations have not supported abolition of nuclear weapons, one contributor to *The Nuclear Turning Point* cast the reluctance into a different light:

[The] residual insurance effect of nuclear weapons against unknown and ominous uncertainties is still appreciated [and the] difficulties of verification and the possibility of breakout are taken very seriously [but the] political will of the nuclear weapon states to live up to their [NPT] commitments is in doubt.

As atomic-pioneer Alvin Weinberg observed, nuclear weapons have reached immortality: Knowledge of making fission and fusion explosives cannot be eradicated. Even so, that does not justify the uncontrolled possession nor use of nuclear weapons. Slavery was once rampant, but its most overt practice was abolished among civilized nations; handguns in some nations are well under control; and weapons of mass casualty have been banned, although transgressions have occurred. Violating an outright prohibition of nuclear weapons would have far more serious consequences; nevertheless, retaliation—assured and overwhelming—remains the ultimate deterrent.

Government fears are captured by the term *breakout*—a covert or overt violation that might disadvantage nations which abandoned their nuclear deterrent. James Schlesinger, former Defense Secretary under President Nixon, considered breakout to be the key danger: "The smaller the [number of] nuclear weapons, the greater is the premium on [an adversary] having just a few nuclear weapons."

To counter fears of breakout, a measured pace is advised: gradualism. The Henry L. Stimson Center (a nonprofit public policy research institution) published a report (signed by former Defense Secretary Robert McNamara, General Charles Horner, and President Reagan's chief arms-control negotiator Paul Nitze) "advocating the elimination of nuclear weapons at the end of a four-stage process." Each stage would be marked by a confidence-gaining pause.

General Butler has summarized and critiqued the case made by nuclear possessionists, to wit: (1) that the weapons cannot be "disinvented," (2) their abolition cannot be verified, and (3) the absence of nuclear weapons will make "major wars" possible. His collective reaction to these arguments is that the "risks of abolition are too often simply asserted as if they could not be adequately mitigated." He takes issue with those who use a risk calculus based on yesterday's or today's sovereign relationships among nations. The general also disputes possessionist reliance on technological tools that now exist, and their addiction to past or current social attitudes. To the contrary, the general asserts that the "stunning reality" of the current global security environment is that it has been "profoundly transformed by the end of the Cold War."

Specifically regarding the assertion that nuclear weapons cannot be "disinvented," General Butler regards that as "merely a truism with no definitive implications for either abolition or retention." This is how he contrasts the two latter end-states:

abolition — "existential deterrence" (all nations are free of the fear of nuclear threats, but have marginal anxieties).
retention — the "nightmare of proliferation" (with the continuing ordeal of a crisis spinning out of control or a dreaded headline that a city has been vaporized).

With respect to verification, General Butler rebukes those who are inclined to spin out "either/or" scenarios: Hardliners argue that if you can't definitively determine whether there are cheaters out there, "you cannot safely go non nuclear." Although absolute verification is not possible, Butler counters that "if militarily significant activity were going on," there would be a "very high probability of being caught," especially in light of the "extraordinary progress" being made in verification.

The third issue tackled by the general is the premise that nuclear weapons preclude a "major war." He faults this "muddled thinking" by hardliners on two specifics: (1) the unstated and false premise "that the Soviet Union was driven by an urge for armed aggression with the West" and (2) the illogical suggestion "that nuclear deterrence was the principal reason why the Soviets did not send their troops [during the Cold War] pouring through the Fulda Gap [attacking West Germany]." Based on access to Soviet archives, Butler has been able to authoritatively contradict those hardliner premises.

On the other hand, Morton Halperin, who is not a possessionist and who has served on the staff of the U.S. National Security Council, advised in 1987 that total elimination of nuclear weapons was not a realistic alternative. He considered verification requirements to be "impossible" to meet at the time in a real-world context.

Despite negative security features of nuclear weapons, nations retain them for many reasons that currently override their deficiencies.

In order to challenge the perceived peacekeeping legitimacy of nuclear weapons, *The Nuclear Turning Point* authors agree that a worldwide network of regional security organizations would have to substitute for unilateral nuclear policekeeping. Reductions in conventional force arms and troops would also have to take place concurrent with abolition.

NATO's Nuclear Posture. Although NATO expansion has already strained U.S.-RF cooperation in nuclear arms control and security, further stress can be averted by reassuring Russia about the purposes of the alliance. According to former president Bill Clinton, prevention of nuclear war should remain paramount in U.S. policy, while maintaining support for "an undivided, democratic and peaceful Europe."

To relieve East-West tensions, NATO could be decoupled from its continued dependency on nuclear weapons: NATO has continued to reserve the right to use nuclear weapons first, whatever the perceived provocation against a member state, even if it's a conventional attack. Because the United States is the world's primary arsenal, it could take the lead in urging revision of NATO's nuclear posture — in particular, abandoning "calculated ambiguity," substituting a no-first-use policy.

According to analysis of Jack Mendelsohn, formerly deputy director of the Arms Control Association:

> The principal threats to the security of NATO and its member-states over the next decades will not come from Russia, but rather from regional dictators, rogue states and violent sub-national groups. The alliance's best defense against these threats is not its nuclear arsenal — the use of which has no military or political justification — but rather its overwhelming conventional military superiority, unsurpassed intelligence gathering and process capabilities and, last but not least, the international non-proliferation regime.

That advice, read many years later, is strikingly prophetic.

Ballistic Missile Defense Deployment. Every nuclear weapon dismantled and destroyed guarantees irrevocable removal of a real threat — one less weapon to defend against. Each liquidation makes a future BMD system more likely to intercept the fewer remaining warheads; thus, the best "defense" against ballistic missiles is disarmament.

BMD development could also benefit from inter-nation cooperative research and development. Pending such cooperation, theater-missile demarcation and ballistic-missile multilateralization programs could be advanced. Also, the theater program could be structured to deal with more imminent threats, while questionable systems need not be prematurely deployed.

Missile-defense negotiation has much to gain from common multilateral interests in reducing dangers from poised weapons. Aside from conspicuous weaknesses of premature deployment, the Arms Control Association has suggested that several counterproductive consequences of striving unilaterally for ballistic-missile defense have already occurred:

> Whatever value one places on a defense against the highly unlikely emergence of a credible rogue-state ICBM threat ... that value pales in comparison with the importance of Russian ratification of START II and the commitment to follow-on negotiations on START III. These treaties would reduce — by about 75 percent from present levels — the number of deployed Russian strategic warheads. In addition, they would eliminate all Russian land-based MIRVed missiles, which are particularly dangerous because their high value as targets encourages a precarious launch-on-warning posture. However unlikely a U.S.-Russian conflict may appear today, Russia remains the only country that can threaten U.S. survival.

Shortcomings inherent in a unilateral missile defense have been amplified by militants have wanted to deploy nuclear-armed missile interceptors. The heavy-handed approach perpetuated U.S. dependence on nuclear explosives, adding another reason why the Bush administration resisted the international norm against testing new nuclear warheads.

Tactical Nuclear Weapons. Although receiving less attention lately, tactical weapons are increasingly relevant because of NATO eastward expansion and Russian hints about their own plans for redeployment. In fact, Russia views sub-strategic weapons "as placeholders of Russian status and prestige in the post-Cold-War world, preventing regional conflicts, and serving as deterrents against strategic escalation."

Some proposed measures call for elimination of naval nuclear tactical weapons (including sea-launched cruise missiles) and all remaining air-launched tactical nukes in Europe, with several hundreds still deployed well into the 21st century.

NATO has suggested transparency measures, including exchanges of data on sub-strategic nuclear forces. Also proposed has been a dialogue on weapons safety and security issues.

An analysis in *Arms Control Today* posited three major incentives for reductions in sub-strategic weapons: possible acquisition by proliferants, potential terrorist acquisition, and susceptibility to unauthorized accidental use. The latter susceptibility is intensified by forward deployment, by sensitivity to communication problems during a crisis, and by decreased controls that might allow use-authorization in the field without the "stringent safety precautions that govern the launch of strategic nuclear weapons."

The United States has reduced its operational tactical nuclear warheads to about 1000, but Russia is believed to have many more. Russia has been balking at tactical-nuclear transparency and at placing their weapons in central storage facilities unless the United States did the same. "The United States and NATO [need to] gain verifiable information about the quantity, security, and safety of these weapons to assess the threat accurately and take steps to prevent their proliferation," but past discussions have broken down because of U.S. insistence on deployment in Europe

Nuclear Materials Insecurity. Efforts to reduce possible nuclear smuggling out of the FSU should get higher priority. The total spent on materials security by the United States before the end of the 21ˢᵗ century was under $3 billion, less than 1 percent of a year's annual military budget. Because these programs are undertaken to reduce visible threats to U.S. national security, they warrant much higher priority than such programs as BMD.

Each of the four basic lines-of-defense being established against smuggling out of the FSU deserves to be beefed up.

Significant and urgent improvements in Russian working conditions and salaries are needed for the guardians of nuclear designs and materials — especially scientists, engineers, and technicians in the closed nuclear cities. At the same time, secrecy has to be loosened enough to increase worldwide confidence in the implementation of measures that reduce nuclear anxiety.

Clinging to Cold War Concepts. Contributors to *The Nuclear Turning Point*, having had relevant experience in and out of government, perceived that "official U.S. policy clings to Cold War concepts." Public opinion, they believed, was sharply divided about the appropriate use of nuclear weapons. Defining "strategy" as the art of matching instruments of national power to the goals of national policy, they found that nuclear weapons and related doctrines were treated variously as "valuable instruments of statecraft," "the foundation of global security," and "useful for deterring a wide range of threats to U.S. interests."

It should come as no surprise that *The Nuclear Turning Point* authors believed such policies to be "fundamentally misguided." Rather, they preferred to see the overriding goal being "to prevent the use, threat of use, or further spread of nuclear weapons." But to carry this through, the United States should "make it clear in both its declared policy and operational doctrine that it possesses nuclear weapons only to deter the use of nuclear weapons by other states." As discussed earlier in this Chapter, strategic and declaratory policy has been going in the opposite direction: toward a broader, overtly politically driven threat-reaction spectrum consisting of both retaliatory and first-use of nuclear weapons.

On the other hand, Alexei Arbatov, who had considerable arms-control experience in the Soviet and Russian governments, warned that "Giving up nuclear weapons without making the world safe for conventional wars or creating too great a risk of nuclear cheating by some state or terrorist group would require enormous sacrifice of national sovereignty by major powers." In highlighting the tradeoffs, Arbatov remembers a joke from Soviet Communist days, recited in the next box.

[**Life Without Nuclear Weapons**]. A man was being tested for joining the Communist party and being instructed about his duties as a good Communist: to stop drinking, smoking, taking bribes, and having mistresses; and to get rid of his excessive living space, extra dachas, and cars.

To everyone's surprise the final demand of the ultimate sacrifice — to be ready, if need be, to give his life for the party — was accepted by him most easily: "What the hell is the sense of such a life" he explained!

Breaking the Deadlock

Americans, joined by others around the world, have become tired of living under and spending for the nuclear sword: According to polls taken at the turn of the new millennium, strong majorities favored dramatic cuts in the U.S. nuclear arsenal, and close to 50 percent endorsed global elimination of nuclear weapons. Nevertheless, Pentagon officials have been authorized to focus on nuclear targeting plans that invite the perpetuation of retaliatory nuclear forces.

Through the remainder of this book, an agenda is presented for breaking the deadlock over significant nuclear-arsenal reductions. Rather than radical changes (which would be unrealistic — although there is nothing wrong with having radical *goals*), a systematic but accelerated stepwise process is suggested that preserves national security and international stability at every stage.

As scientists looking at coordinated reductions, a probability-based approach is suggested; let's call it "nuclear pragmatism." Its endorsement would embrace inclusiveness of nations and weapons, reciprocity in actions, rigid verification at each step, and reasonable time constraints. In order to avoid interim destabilization, nuclear umbrellas and force balances would be maintained without interruption. To ensure that reductions endure, the weapons and their delivery systems are to be dismantled and destroyed, and basic nuclear-material constituents need to be converted into a form that does not allow quick recovery for military purposes.

Nuclear Pragmatism. Physicists — scientists in general, along with other scholars — try to look objectively at a problem and draw conclusions based on relational probabilities. A comparative risk can usually be evaluated from statistical experience with individual events, such as accidents or violations. Without having to determine absolute probabilities, a course of action can be more accurately evaluated if compared with a tangible alternative. This comparative-risk process is a politically neutral means of evaluating relative dangers and benefits of nuclear policies, a pragmatic approach to weapons de-nuclearization tradeoffs.

Natural events and human interactions have random as well as deterministic influences. In making some comparisons, the determinate factors can often be normalized so that stochastic (random) aspects can be separately evaluated and compared. It is during the assignment of various weighting or importance factors that human judgment or preference comes into play.

One qualitative inference that can by drawn from such a conceptual formulation is related to accidental or unauthorized nuclear use: Having a large number of nuclear weapons in arsenals for several years is likely to elevate public risk more than having fewer weapons around for a great many years. Considered in isolation, 10,000 weapons in existence for 10 years create comparatively much greater lifetime-integrated peril (of accidental or unauthorized use) than 100 weapons for a century.

Notwithstanding the simplified calculus of comparative risk, defense hawks prefer the larger number of warheads in order to have more military options during their lifetime. They assign more importance to "national security" than to other risks and consequences.

While elaborate probability equations could be generated, to include other complexities, the outcome would be essentially the same: For large arsenals a dominating risk factor is simply the number of weapons in existence. This rule-of-thumb applies when evaluating ballistic-missile defense effectiveness against an arsenal of many ICBMs: The kill-fraction would have to be unrealistically high in order to have meaningful risk reduction when compared to the alternative of simply reducing the number of offensive weapons.

A sophisticated analysis would take into account the comparative firepower of various types of weapons and delivery systems, but that is unlikely to affect the overall conclusion regarding the lower risk outcome from offensive reductions compared to defensive buildup.

Any political gamble connected with support of a weaker nuclear force is reduced by the military value of retaining a strong, highly redundant retaliatory capability. The latter stability calculus at one time during the Cold War clearly dominated over the dangers of accidental nuclear war and unauthorized detonation.

At some time the number of extent warheads peaked somewhere between 60,000 and 80,000. Qualitatively comparing the chances of nuclear detonations (deliberate, accidental, or unauthorized) against the security benefits of excessive nuclear armaments (deterrence and war-fighting), the insecurity risk clearly has since undergone a significant reduction. Some would argue that accidental/unauthorized dangers — before improved command and control were implemented — greatly overshadowed the security benefits. Later, with the advent of arms control treaties, the detonation risks had become, at best, comparable to the security benefits of nuclear weapons. When arms-control and non-proliferation treaties were implemented, security benefits were high and the nuclear-detonation risk was lowered; the comparative ratio might have become much less than one. Having a low value of the comparative insecurity risk is usually considered a highly desirable public-policy goal.

When the G.W. Bush administration came into office, comparative risk gradually increased, reflecting greater nuclear jeopardy for the public in order to have a more forceful overseas policy. With decreasing security benefits from nuclear arsenals and increasing detonation/proliferation risk, public endangerment became greater. The administration would have disagreed, arguing that non-nuclear risks to U.S. national security improved, thus compensating for the added nuclear risks. Citizens less concerned about overseas political developments were more fearful of nuclear engagement and its possibly unmanageable consequences.

A similar qualitative calculus is implied in the decision-making process for periodic revisions of doomsday-clock settings used by *The Bulletin of the Atomic Scientists*.

Other nuclear-policy determinants s to be taken into consideration are reliability, safety, security, control, and command of these weapons. Psychological factors are important: During the Cold War, a period of political stress, the fear of being vulnerable to a first-strike might have materialized into pressure for a pre-emptive strike. The Cold War "window of vulnerability" was itself a perception based on assessed or assumed capabilities, e.g., missile accuracy and adversary intentions. Each parameter is important: A weakness in one — such as poor administrative and technical control over access and detonation of a single warhead — could significantly alter the total risk factor. These days fewer nuclear weapons means less overall risk as long as conventional forces can provide adequate national security and cope with modern-day threats.

Important risk-mitigation strategies relate to passive-control measures such as launch-abort and self-destruct. Agreements for cooperative improvements in nuclear command and control also help manage the dangers of inadvertent or precipitous reactions. These reduce comparative risk while having little national-security downside.

Also potentially valuable in risk mitigation are active-defense measures such as ballistic-missile-intercept. But the value of defensive tactics is clearly greater when there are fewer weapons with which to cope, in which case a negotiated missile defense becomes more attractive.

Nuclear abolition, disarmament, deep cuts, or prohibition — any of these measures lead to comparative-risk reduction. As nations progress — from emphasis on full-up (fully outfitted) deployed weapons to a more restrained post-Cold-War allocation of disassembled and demilitarized nuclear components — comparative risk to humanity gradually diminishes.

The nuclear pragmatist thus need not get embroiled in the somewhat theological issues about weapons abolition. The pragmatist, while cheering from the ringside, sees considerable value in any singular stage of nuclear reductions.

It is too much to expect the general public or ordinary politicians to fully understand or simply comply with comparative-risk methodology. Rather, policymakers hopefully will appreciate the concept of comparative risk and act upon recommendations that are consistent with a conservative strategy of nuclear pragmatism.

Netpolitik. Whereas logical steps have been suggested to attain goals of reducing worldwide nuclear risk, the question not directly addressed is "How would this come about?" Some negatives seem conspicuous: Certainly it will not happen by itself unilaterally, evidently not by national government self-actualization, and not by international institutions like the United Nations, even if improved.

Assuming a growing body-politic recognizing the mutuality of global nuclear-risk, the experience gained by public individuals and NGOs can be tapped for continued education and pressure.

The enablement of universal networking has been affecting processes of international politics and diplomacy. *Netpolitik* is a name adopted by the Aspen Institute for the "network form as an organizing principle in the conduct of world affairs." While the older term *realpolitik* referred to the advancement of "a nation's political interest through amoral coercion," *netpolitik* "traffics in 'softer' issues such as moral legitimacy, cultural identity, societal values, and public perception."

Control of information has been a major tool of state power, especially in the age of mass media, but modern electronic-communication networks are changing the architecture and facilitation of power and culture. The new technological enablers are no longer peripheral to world affairs but have become a conduit for changes, including values, identities, and social practices. Global networking ignores nation-state boundaries, language barriers, and information controls — thereby furnishing more egalitarian transnational communication to a wider expanse of societal levels.

These capabilities are a powerful engine for newcomers, NGOs, diasporic communities, protesters, commentators, and others to be heard on the global stage. Traditional military and financial powers of dominant nations "are now constrained in new ways by soft power and the politics of credibility." Unilateralism has become an even more problematic policy approach.

The rise of *netpolitik* amplifies the prospects of a common interest in reducing nuclear danger by putting pressure on hardliners who become wedded to outmoded attitudes. As part of a contribution to the dialog, the contents of this book will be updated periodically and noted in the website www.NuclearShadowing.INFO, as well as in associated Google Knols under the author's name.

Freeing the World of Nuclear Weapons (Update 2007). With welcome recantation, some opinion leaders — such as Henry Kissinger and George Shultz, who were part of the nuclear-arms expansion problem when they were in office — have belatedly expressed an interest in freeing the world of nuclear weapons.

In the box that follows is a 2007 French news report of an opinion piece that, partly because of the reputation of its authors (George P. Shultz, William J. Perry, Henry A. Kissinger, and Sam Nunn), seems to have attracted considerable attention.

The reader of this book will find essentially all of these prescriptions accounted for herein, consolidated at the end of this Chapter into a wrap-up about de-emphasizing (de-valuing) nuclear weapons. One must keep in mind that some of the proponents of abolition had helped create a nuclear monster, not so easily willed away.

[Former U.S. Policy Honchos Call For World Free Of Nuclear Arms]. Four top former US foreign policy officials called for a world free of nuclear weapons in an opinion piece.

The Washington heavyweights say the United States should launch a major effort towards banning all nuclear weapons.

Citing nuclear programs in North Korea and Iran, the officials say the world "is now on the precipice of a new and dangerous nuclear era."

Aside from the threat of terrorists using nuclear weapons, "unless urgent new actions are taken, the US soon will be compelled to enter a new nuclear era that will be more precarious, psychologically disorienting, and economically even more costly than was the Cold War deterrence."

In the lengthy article the ex-officials recommended a series of measures that include strong support for the Nuclear Non-Proliferation Treaty and other non-proliferation efforts. But more has to be done, they suggested.

"We believe that a major effort should be launched by the United States to produce a positive answer through concrete stages."

Proposed measures include:

- increasing the launch warning time on deployed nuclear weapons to reduce the danger of an accidental or unauthorized use,

- decreasing the number of nuclear weapons among all nations,

- eliminating short-range nuclear weapons, designed to be deployed with front-line troops,

- providing the highest possible security around the world for all nuclear weapons, weapons-usable plutonium, and highly enriched uranium,

- phasing out the use of highly enriched uranium in civil commerce,

- removing weapons-usable uranium from research facilities around the world.

"Reassertion of the vision of a world free of nuclear weapons and practical measures toward achieving that goal would be, and would be perceived as, a bold initiative consistent with America's moral heritage."

"Without the bold vision, the actions will not be perceived as fair or urgent. Without the actions, the vision will not be perceived as realistic or possible."

[Indeed, any program that would lead to nuclear disarmament has been recognized as having to go through a number of tangible stages that initially reduce ongoing risks associated with the existing nuclear arsenals.

Other prominent individuals and organizations have now jumped on the nuclear-disarmament bandwagon, realizing that the perceived demands of nuclear arsenals for the Cold War have suddenly become a liability in a world where terrorist acquisition of such weapons and their nuclear constituents provide less national security and more public hazard. New levels of awareness and concern have materialized about continued reliance on massive nuclear arsenals, and the globalized nature of terrorism, coupled with recognition that unmanageable expertise and materials from, in particular, the former Soviet Union.]

Iraq Invasion Postscript 4:
Re-Mobilization of Public Protest

The spontaneous and vigorous growth of protest against the 2003 Anglo-American invasion of Iraq is evidence that a significant and extended global population can rally on a grievous public issue. Though the protests were too late and too weak to overcome Bush-administration resolve, a significant public re-mobilization took place — prior to, during, and after the invasion — largely aided by improved channels of communication.

Classic antiwar and peace organizations became revitalized, focus groups formed, and *ad-hoc* alliances coalesced. In fact, an all-time record for simultaneous dissent around the world took place, also shattering the record for protest against any governmental leader in the history of humankind.

NGOs such as MoveOn and the Council for a Livable World were re-energized and facilitated as dot-org accessible, and individuals around the world were enabled by the Internet and wireless communications to speedily access and exchange information about the invasion and its rationale.

Political opposition became emboldened as the alleged weapons of mass destruction were never uncovered, as Saddam Hussein links to Al Qaeda were not proven, and as Iraqi nation-building became more burdensome. The lack of coalition end-game foresight (sensible post-invasion goals, reconstruction plans, and budget) — along with unexpected vigorous and suicidal resistance by unconquered Baathists, Al Qaeda imported into Iraq, and volunteer Islamic Mujahedeen — extended post-invasion public attention and dissidence.

As the rationale put forth by the American and British governments for peremptory action fell apart, newsmedia and public became more aware of the misleading information that had been circulated by governments. Internet correspondence became more incessant about presidential impeachment, opposition-party fund-raising, and more knowledgeable criticism of partisan political actions.

Typical of the disapproval, the Council on Foreign Relations criticized Vice President Dick Cheney's explanation for the invasion in stark but false choices: "They carry listeners from assertions that seem indisputable to conclusions that seem irrefutable. Only on closer inspection, they don't hold water."

Amplifying European public disapproval was an accumulation of other discordant Bush-administration actions: backing away from global-environmental concordance, evincing religiosity and cultural disdain, and slighting collective action against terrorism.

Public involvement in Cold War issues had the indelible result of bringing about memorable lessons to activists and organizers, if not to the Bush administration.

****

Will nuclear arsenals and proliferation be rolled back in the 21 century? The answer depends strongly on many factors, chief among them being universal compliance with the Non-Proliferation Treaty, a cut-off in production of weapons materials, and curtailment of nuclear explosive tests.

Article VI of the NPT embodies unfulfilled obligations of the superpowers and other signatories for movement toward zero nuclear weapons: Each party had undertaken to pursue negotiations in good faith on effective measures to stop the nuclear-arms race at an early date and to move toward nuclear disarmament. Although nuclear elimination might seem to be highly implausible, it is a joint commitment agreed to by the signatories.

Yes, the nuclear arms race has indeed been winding down, but arsenals of nuclear weaponry are still huge. Peak levels were reached with the assertive support of a nuclear-weapons priesthood that influenced information getting to negotiators and policymakers. Post-Cold-War reductions will require viable alternatives. That is one purpose of this book — to communicate nuclear-policy guidance from independent nuclear physicists and engineers, counterbalancing advice proffered by weapons developers and defense analysts.

While circumspection is always justified, especially in matters dealing with national security, now-ancient habitual suspicions and attitudes are no longer warranted. Nuclear-weapon states will not approach deep cuts unless they are to be applied in a synchronized and equalizing manner.

Drawing upon this book's comprehensive technical and political overview — reaching back to the origins of nuclear excesses and spanning forward to the ongoing nuclear impasse— a stepwise comprehensive program of pragmatic, staged nuclear reductions is endorsed. A pace and scale can be chosen to achieve major cutbacks and not endanger stability or security.

In the early years of the new millennium, nuclear-reduction issues were forced to the back burner, as more proximate security concerns dominated world and national events. Many individuals who have recognized the need for comprehensive nuclear reductions were distracted by the imminent national-security implications of Bush-administration foreign and domestic policies.

The remaining three Sections of this Chapter elaborate on **how** *to go about the demilitarization and disposition of surplus nuclear materials (Section E), the verification of reductions in weapons and delivery systems (Section F), and a feasible process for negotiated nuclear disarmament (Section G).*

It's not enough to wish for, or militate for, or decree nuclear reductions. There must be a way to destroy the soul, the heart, the materials from which nuclear explosives are created.

E. DISPOSITION OF NUCLEAR MATERIALS

If deep cuts or nuclear-weapons elimination were to come about, irreversible disposal of current weapons-quality material inventories would be the most enduring undertaking needed to delay resuscitation of the arsenals.

While much of this discussion is necessarily of a technical nature, I'll do my best to make it understandable. Some useful process definitions are reproduced or abridged from Volume 1 and 2.

Surplus nuclear-weapon ingredients have to be disposed of in a manner that is safe, environmentally benign, and — especially in this case — secure, irretrievable, and lasting.

To render nuclear materials militarily innocuous, their disposition would consist of two fundamental stages: dismantlement and demilitarization.

The essential, irreplaceable ingredient of a nuclear weapon is special fissile material composed of highly enriched uranium and/or weapons-grade plutonium; specific high-quality materials are required for all known fission and fusion weapons. Whatever methods for disposal or treatment of high-grade fissile substances are chosen, they should introduce considerable and unavoidable delay—measured in years—to impede rapid reconstitution of nuclear arsenals.

> **[Nuclear Chain Reaction].** To create nuclear explosions, neutron chain reactions are fueled primarily by fissioning fissile isotopes of uranium or plutonium. Although a nuclear reactor can be made to operate in a controlled fashion with natural or slightly enriched fuels (such as uranium), a nuclear explosive requires highly enriched uranium or weapons-grade plutonium.

One measure en-route to irreversible disposal of existing weapons is their mechanical dismantlement. This is an ongoing process, partly because of nuclear treaties already on the books, partly because of revised military objectives, and partly because of normal obsolescence. Even so, no government program yet exists for the systematic demilitarization or disposal of all nuclear materials taken out of weapon cores.

> **[Pits (Nuclear)].** A nuclear core (pit) is a hollow, spherical or aspherical shell of plutonium encased in stainless steel or other metal or alloy, that with its other components makes a fission explosive. In a thermonuclear weapon, the pit's fission-explosive acts as the first (primary) stage in triggering the fusion (secondary) stage.

Most U.S. nuclear weapons are dismantled at the Pantex (Abilene, Texas) or Y12 (Oak Ridge, Tennessee) plants. The basic steps are as follows: After extraction of tritium, the disabled weapon is taken apart and the high-explosive detonator of the warhead is (gingerly) removed. The pit is accessible, and its fissile core is then available for disposition. Salvaged tritium can be recycled to replenish warheads in service.

[Pantex]. Part of my tasking at Argonne National Laboratory was to participate with a DOE team evaluating technologies for unintrusive verification of warhead dismantlement. Among the very sensitive government locations our team visited in the late 1980s was Pantex (Texas) where most U.S. weapons would be taken apart.

The disassembly buildings, called igloos because of their shape, were spaced well apart from each other in case the warhead's conventional high-explosive accidentally detonated — which could and kill everyone one in the igloo and disperse fissionable materials outside.

Thousands of nuclear cores had been assembled in the igloos during the Cold War. Now it was time to dismantle the outmoded ones. We watched in awe as technicians demonstrated the disassembly routine, essentially all manual steps — seemingly blasé about the massive-destruction potential in the small hand-held package.

We understood the precautions taken, such that a fission chain reaction was extremely unlikely during proper dismantlement. But still....

Hundreds — maybe thousands — of warheads were stored in many nearby bunkers. Large sliding doors blocked the entrances. As you walked the aisles, you could hardly escape the disquieting feeling: This one had been targeted for Moscow, that one to wipe out St. Petersburg, this one to annihilate an army unit invading West Germany.

Pantex and Y-12 plants dismantled more than 26,000 warheads from 1975 through 1996. About 4 or 5 could be taken apart in a normal work day — 1000 to 1200 a year — at a pace that is difficult to speed up due to the safety, sensitivity, and specialization of the operations.

If the United States were to reduce its arsenal to 1000 warheads, about 9500 excess nukes would have to be dismantled. Adding the 9500 pits to 12,000 or so already dismantled would result in a huge accumulation of nuclear material awaiting disposal.

As described, Russia too has a well-established framework for storing and dismantling its nuclear devices.

[Nuclear-Weapon Configurations]. Two workable configurations, *gun barrel* and *implosion*, were conceived during the Manhattan Project for making nuclear explosions.

The gun-barrel design, so-called because fissile material was rapidly consolidated by compressing it in a tube, worked only with enriched uranium.

The implosion technique was more complex, requiring inward-directed explosive-driven uniform compression of fissile material. This was the only method that worked with plutonium.

After weapons are disassembled, the nuclear pits can be mechanically demilitarized by crushing or otherwise deforming them. If the materials are simply put in storage or buried, this does not truly demilitarize fissile core material; it could probably be reshaped by mechanical presses into pits within a relatively short time frame. Several processes described below exist by which the fissile materials could be rendered far less recoverable for weapons.

In any event, one must recall that huge enterprises were organized, funded, and operated to produce uranium and plutonium for weapons. These facilities would have to be cut back and eventually shut down too.

[Plutonium Production Rates]. Conversion of uranium to plutonium can roughly estimated based on the power level and duty cycle of reactors. The effective yield (and plutonium quality) is highly dependent on their design and operation.

Under optimal conditions, up to 7 kg/year of weapons-grade plutonium could be achieved in a production-type reactor having a thermal output rating of 20 megawatts. That should be more than enough for a single nuclear weapon each year.

[Enriching Uranium]. Isotopes of natural elements can be separated from each other, *e.g.*, uranium-235 can be extracted out of natural uranium, in a process called *enrichment*, by any of several very complex methods.

Isotope enrichment is very important for weaponization because uranium-235 is *fissile* — that is, when bombarded with neutrons it breaks up (fissions) into smaller components accompanied by the all-important release of nuclear energy and neutrons.

To recap, the first normative stage for disposition of nuclear weapons begins with their disassembly/dismantlement. Demilitarization starts when the pits are mechanically deformed and culminates with irreversible conversion of the fissile materials.

Demilitarizing Nuclear Materials

Demilitarization of nuclear-weapon materials ought to be high on the national and international agenda for the 21st century because they pose a common danger. Weapons-quality plutonium and uranium are the *sine qua non*, the essential ingredients for war-fighting nuclear weapons. Spasmodic reversion (breakout) to a qualitative or quantitative arms race can be obstructed by irreversibly degrading the fissile materials, that is, by demilitarizing them sufficiently to embed long-term barriers against recovery, thwarting their reuse in existing weapons. This durable process contrasts with the alternative of simply storing uranium or plutonium weapon assemblies or components, a passive policy that would retain the potential to reconstitute a nuclear-war-fighting arsenal.

In addition to fissile cores excessed from a post-Cold-War weapons-dismantlement program, large quantities of fissile material in the production pipeline are still in storage. While some is already being downgraded under the aegis of existing programs, more than half of the stockpile had been exempted.

[Weapons-Grade Production Capacity]. Production of weapons-grade materials has long been all but terminated by the major nuclear-weapon states, and capacity for future production of weapons-grade materials is gradually dwindling. One reason is the glut of fissile uranium and plutonium that already exists, partly because new weapons are no longer being produced, and partly because old weapons are being dismantled.

Unfortunately for proliferation management, the development of specialized centrifuge cascades has brought about a means by which uranium can be enriched without the large facilities and electrical power that were dedicated to production in gaseous diffusion plants. Those older (and larger) plants were also more detectable by NTM than the smaller centrifuge facilities.

As for weapons plutonium production, its deliberate creation has been largely curtailed, the primary exception being a few Russian reactors that produce steam and electric power.

While demilitarization of plutonium and uranium need not be part of the early stages of comprehensive nuclear disarmament, it is part of the institutional vector that points to eventual denaturing or elimination of weapons-quality materials.

Isotopic demilitarization does the most to irreversibly impede rapid rearmament. This means changing not only the mechanical and chemical form of fissile weapon materials, but also to degrade their critical isotopic quality. The best candidate facilities to use for converting weapons materials are existing commercial nuclear reactors. Particle-accelerator transmutation and vitrified storage are weak alternatives. These options are discussed below.

[Demilitarization, Degradation, Denaturing]. These three essentially equivalent terms are applicable to fissile materials deliberately rendered unsuitable for use in military-quality weapons.

Demilitarization refers to the intended goal, that is, to render nuclear materials unsuitable for use in a nuclear arsenal. *Degradation* in this book refers to the process or end result of demilitarization. *Denaturing* is the process, as applicable to fissile materials, that leads to their demilitarization; it is not necessarily absolute nor irreversible. Here the terminology is used rather interchangeably.

However termed, the end result represents a major and practical obstacle in re-utilization of fissile materials for nuclear weapons.

Uranium Isotopic Dilution. Fissile uranium can be readily blended with natural or depleted uranium leftovers: This process of isotopic dilution (denaturing) results in demilitarized uranium than cannot be reused for weapons without expensive and attention-getting processes and facilities.

Under existing cooperative programs with Russia, weapon-grade uranium is being blended down to a few percent from its 90+ percent military enrichment. A degraded gaseous product purchased from Russia is processed in the United States and resold in powder form as fuel for existing nuclear reactors. This mode of isotopic dilution is effective in hindering potential military usage of enriched uranium. Also, the intrinsic energy (and economic) value can be recovered — a bargain indeed.

To reclaim its commodity value and to further reduce its potential for restoration as a weapons material, the uranium powder is fabricated into normal reactor fuel used to produce electricity. At the end of 1998 the Cooper Nuclear Power Station in Nebraska was reloaded with fuel rods containing the first uranium that had been taken from dismantled Soviet nuclear weapons. Reactor burnup results in a fissionable residue not usable in military-quality weapons.

Reactor Burnup of Plutonium. Nuclear reactors normally use low-enrichment uranium and low-grade plutonium as fuel. However, reactors can be operated under safe, licensed, and profitable conditions with fuel rods that have higher fissile fractions. All grades of uranium and plutonium can be demilitarized by burnup in reactors.

[Reactor-Degraded Plutonium]. "Reactor-grade plutonium" is usually defined as plutonium containing up to 80% fissile isotopes. Enormous technical difficulties would prevail if a wannabe proliferating nation tried to use such unfavorable material. (Tellingly, none have; all have made nuclear weapons only from pure materials.)

Impure reactor-grade (e.g. *degraded*) plutonium is readily distinguished from the high-quality weapons plutonium used in nuclear weapons.

(Heat and other effects generated by radioactive isotopes in reactor-degraded plutonium introduce all-but-insurmountable problems for warlike weapons — as distinguished from crude, inefficient explosive devices that might conceivably, but illogically, be fashioned out of impure fissile materials.)

The process for demilitarizing large quantities of weapons plutonium is somewhat different from that used with enriched uranium, but the plutonium would end up generating nuclear electricity too. High-grade plutonium can be incorporated into special MOX (mixed-uranium/plutonium oxide) fuel used in the reactors.

[Weapons-Grade Plutonium Processing]. Weapon-grade plutonium metals must be highly refined and treated in a special manner. In order to extract metallic plutonium from reactor-irradiated uranium fuel rods, a complicated chemical, metallurgical, and radiological processing facility is needed. The facility must have capability to separate, purify, consolidate, and convert plutonium to a high-density metallic form. A weapons-plutonium processing facility is not necessarily large, but definitely complex.

While producing energy for the electrical grid, nuclear reactors can simultaneously demilitarize weapons plutonium. Operation with this special fuel composite has been tested and proven in several nations for many years, and most existing reactors — after approved upgrades — can take advantage of such worthy fuel mixtures: metaphorically atoms for peace. Existing government programs in Canada, the United States, France, and Russia have for many years been evaluating and testing the burnup of weapons plutonium.

If a sustained, deliberate program of progressive elimination or deep cuts were implemented, the amount of fissile material available for conversion to reduced potency would increase. To accelerate demilitarization, existing and new power reactors throughout the world could be recruited into a prolonged effort under international controls and safeguards. Such a program would simply be an extension of ongoing efforts and technology to convert present-day surpluses of weapon materials.

[Militarization/Demilitarization of Nuclear Materials]. Despite some confusion among academics, not all grades of plutonium, or even HEU, are suitable for making weapons. As proven from the history and contemporary practices of nations that have made nuclear arms, materials must meet extremely strict specifications before they can be fabricated into explosives of military quality. Only high-grade substances have been employed for that purpose. Nations have supplied their military program by producing plutonium in special reactors or/and enriching uranium to weapons-grade.

The most decisive requirement for fission explosives is purity — isotopic and chemical — of fissile material used to fuel the explosive chain reaction.

Isotopic degradation indeed demilitarizes weapons-grade plutonium — primarily by increasing its fraction of "even" isotopes. The objective would be to meet or surpass the "spent fuel standard" This can be straightforwardly accomplished by means that now exist, such as irradiating weapons plutonium in existing reactors or mixing it with spent fuel.

Based simply on a fundamental understanding of mathematics and reactor physics, one can understand that substituting reactor-degraded plutonium into a precisely machined weapon pit would create a conspicuous dilemma: Either the pit would have to be enlarged, or its essential reactivity available for the explosive chain reaction would be reduced. Furthermore, heat generated by radioactive decay might constitute a serious technical challenge. In any event, a device's military value would be drastically reduced.

To carry out wholesale demilitarization, several things must be accomplished, not the least of which is public acceptance and understanding that there is a solid technical basis for safe and secure conversion of weapons materials. Stringent controls and safeguards over the processing of nuclear materials would have to continue so that the higher-quality fissile supplies can be degraded to levels below their usefulness in weapons. And more civilian power reactors would need to be enlisted for burning MOX.

Thermal-neutron reactors are the foundation of civilian nuclear power. If economical fast-neutron converter/breeder reactors could be built, weapons-grade materials would be consumable at a faster pace, with much more integrated and efficient management of radioactive waste products.

In the distant horizon are fusion-fission hybrid reactors that would be capable of burning MOX or other forms of reprocessed spent fuel. This too has the dual potential of consuming weapons-grade plutonium while burning other transuranics in order to optimize energy production and minimize actinide waste disposal.

[Technical Deductions from Public Sources]. The technical inferences that can be drawn from fundamental, textbook concepts in reactor physics strongly support scientific evidence that sub-grade fissile materials fall short of military-quality standards.

Physicists recognized from the beginning that a highly destructive fission explosive could, in principle, be manufactured from low-grade plutonium. But they also understood the importance of using high-grade plutonium to achieve military-quality weapons, which is exactly what has been done by all nuclear-weapons states.

The high-grade, military-quality plutonium needed for weapons is only produced in special "production" reactors that have brief fuel-burnup cycles. Such short operating intervals are not practical with power reactors, which must run for much longer periods of time to be economical.

Plutonium produced in power reactors, in strong contrast to plutonium-production reactors, automatically acquires a combination of isotopic, radiation, and chemical resistance against unauthorized diversion and illicit use. This is the unavoidable result of long commercially-efficient operating intervals that are easily monitored by international agencies.

Unavoidably, steadily, and irreversibly — because of high burnup — power-reactor plutonium accumulates physical characteristics inherently prohibitive to weapons applications.

Psychological and Institutional Obstacles. Nuclear power — tainted by the specter of Hiroshima, Nagasaki, radiophobia, testing fallout, and Chernobyl — has encountered psychological obstacles. Yet, in the second half of the 20th century, peaceful nuclear applications have proven their constructive value to humanity.

The ultimate in converting swords into plowshares is destroying plutonium by burn up in nuclear reactors, the byproduct being electricity generation. To accomplish it in this century, additional commercial reactors need to be recruited into a dual-purpose role of generating energy while destroying weapons materials. Psychological and ideological resistance to nuclear power would have to diminish. Otherwise, weapon-grade fissile materials will hang around as a long-term temptation.

To overcome perceptional hurdles, it is necessary to transcend non-technical barriers thrown up by misinformed opponents. Some influential arms-control advocates have had a conflicting personal agenda that included putting an end to nuclear power. Often these inclinations have been concealed and rarely disclosed to the public.

While all concerns must be heard, regardless of motivations, it is reasonable to expect naysayers to be held to a substantive scientific standard. For the benefit of those who are uncommitted, this book attempts to highlight and clarify differences in viewpoints. Sadly a hallmark of radiophobia is lack of respect for the scientific method, especially proper characterization of statistical uncertainty. Often statistical incertitude is misrepresented or not expressed.

[Nuclear Power Role in Conversion of Military Stockpiles]. In 2007, 439 reactors worldwide were providing 372 GWe, and 48 new plants were under construction to reach 410 GWe by 2015. More reactors were in planning stages to overcome scheduled reactor retirements. All of this has increased fissile-fuel supply requirements, which are largely based on mining of uranium ore, which itself depends more on the spot price offered than on reserves. Consistent with the current slowdown in the growth rate for nuclear power, there has been a corresponding reduction in the amount of natural uranium being mined.

Although uranium-supply self-sufficiency doesn't exist now, it's not a long-range resource limitation: More uranium mining is likely to happen with higher price incentives, and the use of thorium would make up for any future long-term shortfall. In addition, fast-fission could considerably increase effective fission-energy extraction.

An interim gap in fissile-material supply can be and is satisfied with secondary resources. These are (1) conversion of civilian and military stocks of uranium and plutonium accumulated during the Cold War, including dismantled nuclear warheads, (2) recycling of MOX, a mixture of reactor-grade uranium and plutonium recycled from fuel rods that have undergone a partial burnup/depletion stage in nuclear-power reactors, and (3) re-enrichment of uranium tails.

An exercise of chronic — usually subtle and well disguised — opposition to nuclear power has often been found in the works of a few Princeton University scientists, led by Frank N. von Hippel and Harold A. Feiveson, who have been tireless advocates of arms control. The Princeton group imparted a discernable anti-nuclear-power tone to portions of *The Nuclear Turning Point*, overstating the fear that reactor-degraded plutonium could be used to make military-quality nuclear weapons. (Feiveson has since publicly reversed his opposition to reactor burnup of weapons plutonium.)

Misinformation about weaponizability of reactor-degraded plutonium has been thoroughly challenged: It would be extremely improbable that any nation would be driven to make low-quality explosive devices for military use; this has not happened yet and probably never will. The otherwise commendable *Nuclear Turning Point* has a skewed outlook on reactor-degraded plutonium.

Steadfast convictions have caused a small influential minority of arms-control brethren to distort the proliferation potency of reactor-degraded plutonium. Some fear that the process of consuming plutonium as an arms-control objective might expand the useful role of nuclear reactors; so they would rather forego demilitarization. This negative stance overlooks technical and financial deficiencies inherent in their proposed alternatives, such as vitrified-waste storage or accelerator transmutation.

Much public attention has been drawn to issues about storage and transportation of fissile materials. These are digressions that sidetrack progress in reducing inventories of military-quality weapons. For example, publicized concern about shipping nuclear and radioactive materials through cities has stymied safe and secure storage. If support were given to rail and road bypasses around population centers, these particular difficulties could be alleviated.

Another disconcerting factor has been unnecessary government secrecy, denying information to the public that might help resolve doubts about reactor-degraded plutonium. Failure to release non-sensitive information regarding a 1962 U.S. nuclear test using UK-supplied plutonium is one example. Moreover, DOE classification officials have unwittingly or wittingly blocked the public from having adequate information on military disadvantages of low grades of plutonium. Such misapplication of government secrecy simply muddles future policy on weapons plutonium. Despite directives issued by a previous Secretary of Energy (Hazel O'Leary), her government-openness initiative has penetrated only skin deep in the bureaucracy.

[MOX Conversion Facility in South Carolina (2009)]. Plutonium and nuclear-material management missions, now conducted at the DOE Savannah River Site, are being expanded to include materials taken from dismantled U.S. weapons and from other DOE sites. This new mission focuses on the disposition of excess weapons-grade material consistent with the U.S.-Russian agreement on nonproliferation.

Excess weapons-usable plutonium will be converted to a (MOX) form that can be used in commercial power reactors. Construction began in August 2007 for a mixed-oxide fuel-fabrication facility. Two foreign companies (Mitsubishi, Japan, and Areva, France) have jointly invested in and will be operating the dedicated civilian nuclear-fuel-fabrication facility.

A separate DOE facility at the site will disassemble nuclear pits (cores) taken from U.S. weapons.

In addition, if needless bilateral secrecy with Russia could be reduced, reciprocal verification of nuclear disarmament would be far less expensive. Nuclear materials in storage could then be protected based on existing security needs rather than prior significance. Sufficient data could be released to fashion a domestic and international safeguards policy that relies on cost-effective analysis, not scare tactics. The expenses of safeguarding fissile materials could be held in check by applying appropriate measures proportionate to realistic diversion risk.

The preceding paragraphs have mentioned some implications of government secrecy and headstrong opposition. Other forfeitures ensue from ideology-based reluctance against moving directly toward elimination — rather than storage — of surplus nuclear-weapons materials. Some members of the arms-control community are their own best enemies.

Because the United States retains a formidable stockpile of stable, certified warheads, preservation of capability for massive Cold-War-scale nuclear rearmament is no longer needed. Funding that promotes nuclear weapons continues to bloat the DOE defense-program budget.

A studied transition toward fissile-material degradation would be a wise option for successive government administrations. The burden of long-term nuclear-materials disposal inevitably falls upon future generations. Methods chosen now ought to be based on a cost-benefit evaluation which recognizes that different grades of plutonium and uranium can be demilitarized according to their respective risks — real, rather than hyped.

Tritium

The special hydrogen-like radioactive isotope called tritium, mentioned frequently in this Chapter, is a central ingredient in nuclear weapons that use fusion reactions to boost the explosive yield. Gaseous or liquified tritium is incorporated in boosted-fission and staged-fission-fusion devices. Because tritium naturally decays at the rate of 5.5 percent/year (a factor of two loss every 12 years), boosted-weapon gas reservoirs would need to be refilled every few years.

Tritium is created by special processes in nuclear reactors or accelerators. Some estimates are that the United States had produced the equivalent to 225 kilograms of liquid tritium through 1988, when production was stopped. At the millennium only 75 kilograms remained (and boosters are estimated to require about 4 grams each).

[Boosted Weapons]. Fission weapons are said to be "boosted" if they have an enhanced yield due to the presence of gaseous deuterium and/or tritium, which support thermonuclear fusion reactions. Fusion reactions contribute to the explosive energy by effectively magnifying the fission rate. For a given yield, fusion-boosted devices can be made somewhat smaller and lighter than fission weapons.

If nuclear-weapon stockpiles were to be reduced at the rate of approximately 5 to 6 percent/year, either by planned obsolescence or by arms-control measures, little or no tritium production would be needed. The quantity of tritium required to maintain the nuclear stockpile depends on current inventory (in deployed and reserve weapons) and on projected arsenal reductions, taking into account losses due to radioactive decay.

[Fusion Weapons]. Fusion (thermonuclear, or hydrogen) weapons, are nuclear devices in which an uncontrolled, self-sustaining fusion reaction takes place — resembling the large-scale nuclear reactions that power the sun. In a fusion reaction, two energy-rich nuclei collide and combine ("fuse"), rearranging their neutrons and protons into reaction products and releasing energy.

Thermonuclear weapons are multistage devices quite different from fission and fusion-boosted bombs, and far more destructive. They contain three essential components or stages: (1) a fission-based primary "trigger," (2) separated fusion fuel (such as lithium deuteride), and (3) a massive casing. Each stage is important in obtaining a high explosive yield. Typical multistage fusion weapons have yields in megatons.

The U.S. government has considered a number of options: restarting the Savannah River production facility (estimated to cost $2 billion), extraction of tritium from modified civilian power reactors, or construction of a dedicated tritium-production reactor. Although all interim requirements could be filled by foreign sources of tritium, the United States has been reluctant to depend on them. Canada and India have heavy-water power reactors that produce tritium as a saleable byproduct.

Before the new millennium, DOE said that its tritium supply would last until 2010 at START I levels, which allowed about 8400 deployed warheads. (In addition, DOE has several thousand reserve warheads loaded with tritium that could be tapped.) If Russia and the United States had decreased their strategic warheads by a factor of about two under START II, new tritium would not have been needed for many more years.

If START III were to trim the number of warheads by another factor of two, this would add a decade of delay before having to reconstitute production of tritium.

Together, these measures could have postponed the tritium resupply problem to 2020 at the earliest. However, in 2003, DOE started to produce tritium at the TVA Watts Bar nuclear power station, breaking away from the nonproliferation-restraint policy of keeping separate the peaceful and weapons functions of civilian reactors. DOE had determined that using commercial American reactors was (1) more flexible and cost-effective than constructing a dedicated tritium-production reactor, (2) would abstain from dependence on foreign sources, and (3) would "avoid tapping into its five-year tritium reserve."

[Tritium-Production Update (2009)]. One of the first steps taken by the Obama administration, in connection with its required Nuclear Posture Review, was to reduce tritium production because projected nuclear-mission requirements have decreased. Tritium was being produced at the Watts Bar TVA nuclear reactor and extracted at the Savannah River DOE site.

Burial of Nuclear Materials

While underground burial of nuclear materials might satisfy those who oppose MOX reactor burnup, the most durable measures to solidify deep cuts comes from the actual destruction of weapons materials. Burying them leaves open the prospect of recovery, whereas MOX burnup in reactors irreversibly destroys the military value of uranium and plutonium. Besides, technical feasibility for safe and economic underground storage of weapons-grade materials is elusive.

If a deep-cuts program became enacted, especially abolition of nuclear weapons, huge quantities of nuclear materials would have to be demilitarized. While the problems are similar to those posed at present, they would be aggravated by the larger quantities to be processed and by the greater need for assured irreversibility.

Underground burial of weapon-grade materials is riddled with unresolved policy and technical issues. It is surprising that arms-control advocates would favor it. Because of the diversionary aspects of that controversy, research on the selective burial of heavy elements with long half-lives has been postponed. Also needed is public education about realistic radiation standards and exposure. If sequestration (burial) of weapons-grade nuclear materials were to move ahead, underground storage of radioactive waste would have to be accepted.

A forward-moving step could be to designate the Yucca Mountain repository for temporary (i.e., 200-year) storage of spent reactor fuel. After weapons-grade plutonium is consumed in reactors, depleted fuel would eventually need to be stored underground. Approval and operation of a storage site in Nevada would become a national-security issue. Its operation would also begin to relieve the congestion of traditional spent fuel temporarily kept at surface-level utility sites.

Military nuclear waste (from producing weapons-grade plutonium and uranium) is sequestered by the U.S. government in a converted salt mine near Carlsbad, New Mexico.

A logical national-security role exists for burial of immobilized demilitarization byproducts, limited simply to low-grade fissile and radioactive materials that are surplused from nuclear arsenals. Meanwhile, excess weapons-grade plutonium can be transformed to a more irreversible substance by MOX burnup in reactors.

[National Academy Report on Yucca Mountain]. In its 1995 report, "Technical Bases for Yucca Mountain Standards," the National Academy of Sciences (NAS) recommended that compliance with EPA surface-radiation standards be based on an estimated time of public-radiation peak risk, "within the limits imposed by the long-term stability of the geologic environment, which is on the order of one million years." The Academy asserted that calculations for Yucca Mountain show "peak risks might occur tens- to hundreds-of-thousands of years or even farther into the future."

Frankly, it is inconceivable to an experienced engineer that any standard could be promulgated beyond existing industrial experience.

The NAS did offer a caveat: "The situation to be avoided ... is an extreme case defined by unreasonable assumptions regarding the factors affecting dose and risk.... We conclude that an individual-risk standard would protect public health, given the particular characteristics of the site, provided that policy makers and the public are prepared to accept that very low radiation doses pose a negligibly small risk."

Nevertheless, the Courts and the federal agencies have made exaggerated assumptions about dose and risk dangers both in the near term and the long term.

The ambiguous and contradictory NAS report language has contributed to misunderstandings. For example, the report says "We believe that compliance assessment is feasible for most physical and geologic aspects of repository performance on the times scale [in the order of a million years]." Yet the NAS admitted, "It is not possible to predict on the basis of scientific analyses the societal factors required for an exposure scenario."

By "cherry picking" desirable quotes from the report, one can support almost any case for or against licensing of Yucca Mountain depository.

Absent substantive justification to the contrary, radioactive-storage facility life-time standards could be reasonably and simply met by eminent-domain purchases of ranches within 18 kilometers of the desolate site so that long-term human trespass can be prevented.

Cost of Disposition

The 1996 *Atomic Audit* estimated that storage and disposal costs of surplus fissile materials from retired nuclear weapons were in the range of $15-25 billion added to the $11 billion already spent.

To reduce the U.S. arsenal to about 10,000 warheads would require retiring and dismantling another 12,800 warheads stored in the states of Louisiana, North Dakota, Georgia, New Mexico, Nevada, Missouri, Virginia, and Washington. In 1992 the Department of Defense began stowing nuclear weapons at the Kirtland (New Mexico) Underground Munitions Storage Complex.

The cost of denaturing uranium is largely recovered by reusing the lower-grade product as fuel in existing nuclear reactors. That's the U.S. Uranium Enrichment Corporation's goal for surplus Russian enriched uranium.

Weapons-plutonium conversion costs could be offset if MOX reactor fuel were fabricated and sold to civilian reactors; it's not clear if this would become profitable, but it would certainly defray a significant portion of the investment while satisfying a universal arms-control objective.

[Yucca Mountain Vicissitudes (2009)]. Licensing and construction of the Yucca Mountain nuclear-waste storage facility has wavered because of changes in regulatory requirements, legal disputes, and political vacillation. The facility was designed to store a backlog of radioactive waste, civilian and military, accumulating at more than 43 locations.

In 2002, President Bush and Congress approved Yucca Mountain as the federal repository for used (spent) nuclear fuel and high-level radioactive defense waste. The underground repository was designed to hold 70,000 metric tons of waste. About 10 percent of the waste would come from government defense programs, including naval reactors.

DOE had been in a lengthy process of applying to the Nuclear Regulatory Commission for a license to build the storage facility.

Soon after taking office in 2009, President Obama's administration issued a statement indicating that it did not want to fund the project. No alternative was proposed for the $10 billion investment already funded by nuclear utilities and the government, thus leaving nuclear waste in limbo, stored at reactor sites throughout the nation.

For denaturing plutonium, the monetary burden can be offset by producing consumer-marketed electricity, district heat, desalinated water, and hydrogen fuel. These are all environmentally benign products.

The plutonium-demilitarization process, thus the cost, can be greatly expedited by using fast-spectrum reactors, like the BN-800 under construction in Russia. One American firm has proposed an intermediate-neutron-spectrum reactor, using molten-salt carrier/coolant in the core.

Simply stated, demilitarization of weapons-grade nuclear materials, besides being beneficial for national security, is a money-making proposition.

****

Huge contemporary inventories (a great many tons) of weapons-grade plutonium add to the risk or consequences of proliferation and terrorism. But the technology and process exists to demilitarize it: With reactor irradiation and burnup, these materials would gain (additional) protective radiation and isotopic barriers that become intrinsic. The means and the incentives already exist to demilitarize uranium.

The unsavory alternative is to keep these dangerous materials in storage throughout the world without any reduction in their weaponizability.

Nuclear technical policy can support several national goals. One of these is to isotopically demilitarize plutonium so that it cannot be quickly put back in weapons. Another is to safeguard materials with high security and safety, while disposing of plutonium in a sensible and economical manner. The technical feasibility for plutonium (and uranium) degradation and the ability to verify the conversion are on sound technical ground.

An egregious form of resistance to plutonium demilitarization has been presented by vocal opponents who almost inevitably have asserted technical positions without abiding by scientific standards of statistical uncertainty. The disputants have often made assertions that imply far more confidence than warranted. This is particularly the case for well-meaning individuals who have had much academic but little laboratory, field, or experimental experience.

[Regulatory Actions Regarding Yucca Mountain]. In its 2001 rulemaking, EPA noted that it was not possible to make reliable estimates of the repository's performance over long time frames. In the face of these uncertainties, EPA adopted a 10,000-year compliance period, consistent with the Waste Isolation Pilot Plant in New Mexico and many international geologic disposal programs. EPA considered peak-dose estimates recommended by the NAS as a valuable, generalized indicator of disposal-system performance. DOE was required to include them in its environmental impact statement.

Before the Yucca Mountain repository could open and accept waste, DOE had to demonstrate that it could meet EPA standards both under normal conditions and in the unlikely event of "human intrusion" — if actions such as drilling for water or other resources breached the waste containers. In such situations, the public must not be exposed to more than 15 millirem of radiation per year up to 10,000 years and not more than 350 millirem between 10,000 and 1 million years.

However, the U.S. Court of Appeals in 2004 decided that the 10,000-year regulatory period was not "based upon and consistent with" the NAS report's recommendation and struck down that portion of EPA's Yucca Mountain standards.

Otherwise, the Court validated the approval of Yucca Mountain rejecting all legal challenges to the project except the one above.

The primary mode of radioactive-waste shipment to the site was to be secure rail, including new tracks where possible. Although nuclear waste was not likely to be ready until 2012, the site could have reached its statutory capacity by then because of unmet national storage needs over the preceding decades.

For nations to benefit from nuclear demilitarization, several cognitive things must happen too: Vested interests need to be disclosed or revealed, especially those that hide behind an official cloak of secrecy. Ideologically motivated opposition is in need of enlightenment through rational ("dianoetic") reasoning. Finishing the task of controlling and eliminating nuclear arsenals will have to be given high priority.

Twenty-first century progress in nuclear-weapons elimination will probably be incremental, the first step being demilitarization of twentieth-century weapon-grade materials.

F. VERIFICATION ISSUES

In Volumes 1 and 2, innumerable situations are described from before and during the Cold War where verification was an essential and facilitating concomitant to arms agreements. Avoidance of treaty verification signified underlying political discord more than functional difficulties or soundness of the procedural and technical means available to validate compliance.

Simply put, verification can be and has been crafted to provide continuous assurance against militarily significant deviations from explicit agreements.

Substantiating compliance with treaties is considered an essential element of arms limitations among adversaries or former adversaries. There would be little prospect of deep reductions if nations believed their security could be undermined by undetected violations. Of course, military planners would configure residual forces so that deterrence would not be sensitive to undetected cheating and so that compliance validation procedures could, in a timely and convincing manner, detect cheating or preparations for breakout. This process is consistent with the inherent conservatism of military planning which is, or should be, based on updated design-basis threats to national security.

Perhaps of greatest importance are the verification lessons learned whenever innumerable obstacles had to be overcome, such as: devastating arsenals, big stakes in the outcome, mutual and indelible suspicion, vested domestic interests, inadequate compliance technologies, public-relations campaigns, and persistent reluctance to accept results. Verification was sometimes negotiated in a step-by-step fashion, codified with explicit directions, institutionalized with designated responsibilities and mechanisms, reinforced with treaty-implemented technology and procedures, backed up by national technical means and abiding diligence, and presented as a package to the public and to legislative ratifiers.

> *While treaty negotiators settle on means of verification suitable for specific objectives, readers will find here an analysis of explicit or implicit recommendations, especially for treaties that have been under consideration.*

The Nuclear Turning Point authors described in 1999 a comprehensive means for systematic nuclear reductions, to be carried out by a multipartite verification authority having the following major responsibilities:

1. Monitoring agreed numbers, production, and stockpiles of warheads and fissile material.
2. Verifying allowed numbers and types of nuclear-delivery vehicles.
3. Confirming restrictions on deployment and alert status of permitted nuclear forces.

Verification can be accomplished by using unilateral means or cooperative arrangements. For treaties involving many parties, special international agencies might have responsibility, as is the case of the NPT, where the IAEA carries out its own inspections and verification of facilities under international safeguards.

The On-Site Inspection Agency (OSIA) has been responsible for most U.S. negotiated verification, including implementation of the INF Treaty. ACDA had primary responsibility for assessing verifiability of proposed treaties. Interagency rivalries and turf battles in the United States have historically hampered and preoccupied delegations and negotiations on verification.

Other nations have their own chartered inspection teams. Self-reliant national means (with some assist from international or multinational agencies, would continue as part-and-parcel of the verification package.

Treaty breakout is the design-basis "threat" usually considered in crafting a model for verifying deep cuts. If nuclear weapons were reduced to the very minimum, the gravest perceived threat to mutual confidence would be a potential adversary's residual capability to reconstitute an offensive nuclear force. This potential is especially intimidating if there's a track record of having previously created an arsenal of offensive-quality weapons that used proven designs, materials, and technologies to fabricate a military-service-certified weapon. In short, the original five nuclear-weapons nations have an inherent advantage on rearmament. Similarly, the other smaller nuclear states have a jump up on those that have never weaponized.

Too much attention — especially from some individuals and NGOs — has been devoted to an overstated potential for manufacturing nuclear weapons from low-quality fissionable materials; such devices couldn't be an offensive threat. One upshot is that policymakers have been diverted from the transcending need to control existing arsenals of nuclear weapons and weapon-grade materials.

As can be noted by scrutinizing the Consolidated Table of Contents at the end of this Volume, considerable history and experience with treaty verification has been included in Volumes 1 and 2 of *Nuclear Insights*. Lessons learned as well as lessons buried are recounted for the purpose of providing future guidance.

Standards of Compliance, Means of Verification

Standards of compliance and the means of verification have a better chance of agreement among parties when venomous rhetoric is avoided. "Military significance" better serves as a useful standard in place of the vague terms "effective" and "adequate" applied in the past.

No verification can be "absolute" — nothing like that exists reproducibly in life. However, high probabilities can be expected in the judgment process for evaluating verification and intelligence data. Military significance is a value judgment that follows from the process, which includes designated professionals making decisions based on the sum-total of information available, as well as affairs of state. Here are some factors taken into account.

On-Site Inspection. On-site inspection and cooperative monitoring processes came into their own with the INF Treaty, under which it was demonstrated that missile-production plants could be monitored on a continuous and non-intrusive basis. A 1988 Joint Verification Experiment between the United States and the Soviet Union showed that instruments for hydrodynamic and seismic measurements could be exchanged and used at each other's sites to verify the agreed limit on explosive yield.

Usually items subject to treaty limitation (TLIs) are designated to be under particular scrutiny, as in START I and II. In fact, the limits of verification for a given TLI could well be preset at a matching sensitivity: not too coarse — to avoid undercounting—nor too detailed—to reduce unnecessary information.

The history of inspections for treaty verification, going at least back to the aftermath of World War I, is particularly relevant.

Managed Access. The inspected state can manage inspector access in order to protect sensitive installations and locations. Access management provides a functional balance between the need — in the case of a CTBT — to see whether a prohibited test has taken place, while still protecting sensitive information. The host country is able to negotiate the manner, and thus the intrusiveness, of reciprocal inspections at sensitive facilities. As a result, activities not demonstrably relevant to the goals of a treaty at military bases and commercially sensitive facilities (or parts thereof) would to avoid alleged military and industrial "espionage" conducted under the guise of verification.

Open Skies. Based on the example set after successful negotiations of the CFE agreement, widened application of cooperative monitoring has proven to be feasible. For example, the Open Skies Treaty depends on individual nations allowing aircraft outfitted with agreed instrumentation to reciprocally monitor territorial activities and limitations (at a resolution capable of distinguishing trucks from tanks). The aircraft are outfitted with side-looking synthetic-aperture radar that can differentiate large military equipment through clouds and at nighttime. Other sensors allowed have been optical, video, and infra-red cameras.

The Open Skies monitoring arrangements have been routinely carried out, with reciprocal overflights taking place in Europe and North America.

Joint Consultations. Of great value to the successful implementation of START I and II, especially during contentious times in the past, is the Standing Consultative Commission. This forum was created to enable resolution of day-to-day problems regarding treaty implementation. Because consultations between the treaty parties were held in (classified) meetings closed to outsiders, public posturing was muted during those controversial periods.

Reciprocity. Washington, insistent that stringent on-site inspection be required for most arms-control negotiations, has been quite reluctant to actually accept and implement such provisions when it came to inspections of U.S. facilities. For instance, to implement the CWC, the United States passed domestic legislation adding conditions that effectively blocked unlimited access to U.S. chemical facilities. Similar unwillingness to accept reciprocity has hampered agreement on verification provisions for the BWC.

Proportionate Measures. It requires more intrusion to identify traces of a hidden biological weapons (BW) program than it does to detect evidence of an underground nuclear test. For a given treaty, intrusiveness depends on the nature of the objects or substances being regulated. Allowed verification measures are expected to be proportional to their treaty significance and to the overall military balance of power. Introduction of measures disproportionate to military significance would be opposed.

The G.W. Bush administration, displayed deep-felt institutional hostility to arms control and reciprocal verification, countering with alternatives that required no inspection or validation of BWC compliance. The administration suggested the following alternatives: enact national criminal legislation to criminalize activities prohibited by the BWC, adopt strict standards for security of pathogenic microorganisms, establish a mechanism for international investigations of suspicious incidents, set up voluntary cooperative mechanisms for resolving compliance concerns, and implement strict biosafety procedures. While the foregoing are useful adjuncts to a treaty, none of these proposed provisions would require nations to submit to outside compliance verification.

National Technical Means

Continued improvements in overhead and electronic surveillance technology have meant significant advances in NTM — national technical means. These advancements supplement formal verification, lessen dependence on inspections, and reduce the potential for military surprise.

With a half century of intense and generously funded intelligence activities, some governments have been able to develop information-collection methods for all the physical media — underground, surface, air, and space — as well as all usable physical phenomena — e.g., sound, light, electromagnetic and seismic waves, radiation, and heat.

Commercial satellites provide a useful exemplar. The *Ikonos* high-resolution visible-band imaging satellite was "able to capture one-meter resolution of nearly any object on the earth's surface." Although not able to distinguish people, *Ikonos* could depict a launch pad at North Korea's Nodong missile site. Using images purchased from the private company Space Imaging, the FAS deduced that the North Korean long-range missile program was not nearly as advanced then as many U.S. officials made it out to be. (Because of such independent evidence, hardliners like Frank Gaffney had to soften their oratory from "grave harm" to the "potential" of the North Korean program at the time. The rockets were more of a regional threat to neighboring nations. (Since then, North Korea has continued to test and improve medium-range ballistic missiles.)

Earth-orbiting satellites can be tuned to detect military vehicles, installations, production plants, and other facilities. Few human-made or natural outdoor activities can escape observation, especially by the high resolution and broad bandwidth of military and intelligence satellites. Underground activities usually leave at least tell-tale visible clues.

In the course of time, more rapid access to high-quality, real-time data has become possible. Major governments are now in a position to derive significant independent and shared information needed for their strategic security. And there are always spies, defectors, and whistleblowers.

Because of an increase in worldwide travel and communications, unilateral and cooperative verification has a far greater reach now. With the exception of a few relatively isolated nations, travel within borders has become more frequent; civilian aircraft fly in and out; and communications over wire, cable, or radio frequencies are much more common. All of these improve opportunities for human intelligence and NTM, including planted "bugs," surreptitiously mounted sensors, and monitored radio waves or cables.

Naturally there are physical and financial limitations on NTM: Underground or indoor activities are usually not subject to direct scrutiny. If a warhead or weapon were being moved from one place to another above ground, NTM most likely will not be able to distinguish between nuclear and conventional devices.

In any event, threatening activities that might have strategic significance are now much easier to monitor by NTM, while activities of local or regional impact still require supplemental means of clarification.

How Many, How Much?

When it comes to verification of deep cuts, two basic questions present themselves: How many nuclear weapons are there, and how much nuclear material still exists? These questions presuppose that reliable numbers exist on how many weapons were manufactured and dismantled, and how much material was originally produced.

Accountability for weapons-delivery systems is somewhat straightforward, mostly due to their larger size; so the focus in this Section is on the more difficult challenge of verifying the number of warheads. Nevertheless, an edifice of comprehensive verification would include delivery systems and other means of production.

The objects of nuclear accountability could be either the fissile constituents or the warheads themselves — or some combination of both.

Even if data were exchanged on the quantity of fissile material that had been produced, it would be difficult to create a materials-balance that would precisely account for the number of warheads originally manufactured and dismantled. In the haste of early production, fissile material was produced or discarded without accurate accounting, and some was trapped in unaccountable configurations of pipes and tanks. Another difficulty might come from the sensitive nature of the information that would have to be divulged, especially if international inspectors are involved.

Conversely, if warheads—rather than fissile materials—were to be the sole unit of accountability, uncertainties would still remain because documentation and independent validation might not exist to adequately verify history of manufacture or dismantlement. Yet, warhead accountability, supported by the registration of specific data, would be central to the verification of deep cuts.

In order for a database submitted to appropriate verification authorities to have credibility, certified historical records of nuclear-materials production and disposition would be a valuable supplement because supportive inferences could be drawn from indirect data or auxiliary measurements.

Fissile-material production records could be partially verified by using facility operating-history records and measured parameters that would at least outline the boundaries of uncertainty in warhead declarations.

Keeping track of warheads by means that are dependable but not overly intrusive is important. A unique-signature identifier pattern can be recorded for a small surface area of the warhead outer casing. With such a surface-area "fingerprint" (see below) the weapons could be reliably but nonintrusively tracked from their in-service location all the way through the steps involved in their dismantlement.

At the earliest stages of treaty implementation, it would increase confidence if weapons-signature data were registered and verifiable inventory declarations were recorded. This registry, like other disclosed data, would be validated when, or soon after, a reduction treaty comes in force. Of particular interest (agrees *The Nuclear Turning Point*) would be "a comprehensive declaration of the location, type, status, and unique identifier for all nuclear explosive devices and canisters containing pits or other forms of fissile material." These declarations would apply to every form of deployment—land, sea, and mobile—or storage of nuclear weapons and nuclear components.

With such measures and given sufficient access, any movement or change in weapons status, and any allowed new (*vis-à-vis* historical) permitted production of fissile materials, would comparatively easier to monitor with confidence.

Warhead Disassembly

Because of its crucial role, verification of nuclear-weapons disassembly is exceedingly important. Unfortunately, cradle-to-grave accountability for all the weapons produced will not be possible, partly because production records and materials tracking are incomplete and partly because mutual transparency is yet to be instituted. This is a consequence of more than a half-century of irregular national record keeping, changes in governments, the absence of cooperative data sharing, residual adherence to secrecy, and continued reluctance to disturb "sleeping dogs." Nevertheless, as the number of warheads remaining in service gets smaller, the importance of warhead-disassembly verification increases.

> **[Dismantling U.S. Nuclear Warheads].** Warheads removed from the U.S. stockpile have been dismantled on a three-shift five-day work week at the Pantex Plant in Texas.
>
> From 1945 to 1990, the United States produced approximately 70,000 nuclear weapons of roughly 70 types for more than 120 weapon systems. The United States has dismantled more than 60,000 warheads. When the Cold War ended, there were approximately 21,500 in the U.S. stockpile. More than 11,000 were disassembled and disposed of during the 1990s, leaving about 10,400 nuclear warheads in the stockpile.
>
> Disassembly is done in one of 13 cells known as "Gravel Gerties," reinforced rooms able to withstand an explosion equivalent to 250 kilograms of TNT.

Transparency and irreversibility, aimed at full chain-of-custody for declared warheads, have been the goal of unconsummated UN Conference on Disarmament negotiations. Closure in the entire accountancy loop from weapons production to disposition would require, at the least, detailed historical records to check on internal and external consistency.

Verification of warhead withdrawal from service is a complementary phase in the overall process of ensuring that real nuclear warheads are removed from delivery systems. The chain-of-custody can usually be initiated at that point, ending only after dismantlement. This is where tagging of warheads and RVs becomes useful.

How would one find out if the item is a real warhead, not a dummy? This can be accomplished in part by means that relate to deployment. If RV removal from an operational missile is witnessed by inspectors, strong credibility is attributable to the warhead's validity as a nuclear explosive. After all, any object — a warhead, dummy, or penaid — installed on a missile is accountable as a warhead unless confirmed by external measurements to be otherwise.

When the missile shroud is offloaded in the presence of inspectors, the obligatory chain-of-custody would begin for the RV and its warhead(s).

For aerial bombs and tactical weapons, other agreed measured characteristics would have to be used to verify the warhead as a nuclear device. Some of these capabilities have been described elsewhere in the three volumes of this book.

The ultimate means of supplementary verification would be to have "anywhere, anytime" access for inspection, which would be particularly needed as the number of warheads is reduced to levels of, or below, ~200 apiece. Because nuclear warheads have a finite service life, guarding against breakout necessitates inspections at facilities that might be used to replenish and recondition the warheads.

Weapon labs and their unremitting supporters are among the more entrenched constituencies likely to oppose inspections under a multilateral agreement to reduce warhead inventories. In national-security circles there has been considerable (exaggerated) fear of having inspectors from other nations gain access to nuclear-weapons facilities, even though meaningful information would be very difficult to discern from external features.

[**Warhead Dismantlement**]. The process of warhead dismantlement follows these generic steps:

At the end of a warhead's warranty life, the military service to which a warhead is assigned returns it to the responsible government organization that ships the warhead to a storage facility collocated with a dismantlement plant.

The first step in the warhead dismantlement process is separation of its "physics package" (nuclear explosive) from the rest of the warhead. After that, the warhead "primary" is separated and taken apart. The disassembly of its primary begins with the removal of its outer casing, electric blankets, and its firing system (detonators and cables). Then, plant technicians remove the primary's high-explosive components to gain access to the [fissile] metallic core, which is disassembled into its tamper, reflector, and "pit" components. The fissile material pieces are placed in storage containers.

Thermonuclear "secondaries" go through a different pathway for dismantlement and materials recovery, possibly at a different facility.

Chemical and radioactive wastes generated in the course of dismantlement are stabilized, compacted, and disposed. Warhead components and subassemblies are transferred for further disassembly and storage or disposition sometimes at the corresponding manufacturing facilities.

Unique Identifiers. In *The Nuclear Turning Point* it is recognized that "Verification would be improved if all declared nuclear warheads and canisters containing pits or fissile materials were equipped with a unique identification number or tag that was specified in the declaration." The identifiers help maintain chain-of-custody, reduce intrusiveness, and allow inspections to be made randomly — thus reducing inspection effort and interference with normal plant operations. Discovery of any untagged item would immediately constitute an anomaly that would require satisfactory explanation.

Unique identifiers could expedite and reinforce routine and short-notice inspections at declared facilities where warheads are authorized. Accountable warheads that are already identified would not have to be reexamined. Warheads not so marked could be subjected to non-intrusive evaluation (e.g., radiation metrology), thus greatly reducing vulnerability of militarily significant information.

Intrinsic-feature identifiers rely on existing serial numbers and surface features (as described earlier). Tamper-resistant tags (such as bar-coded labels or holographic images) could be affixed. Tamper-resistant seals fasten a closure in such a way that its unauthorized reopening would be readily recognized. In any case, to assure reliability, physical access is required for installation and verification.

When RVs are removed at missiles sites, only shipping/storage containers need be sealed, but it would be preferable to tag both the warhead and the container, if feasible. Nuclear bombs removed from aircraft or storage sites could also be validated and given the same traceability. Tagging the warhead itself had the advantage of making it verifiable through all stages of transit, storage, and dismantlement.

Unique surface-feature identification is one way of keeping track of warheads from their removal from service through ultimate dismantlement. For example, all warheads probably have (can have) some inscribed indelible designator on their outer casing. These are permanent etched or mechanically inscribed markings normally used in maintaining routine inventory and custodial control of the weapons.

Such markings are subject to being "fingerprinted" by a tamper-resistant process. The three-dimensional features of the marking can be faithfully and simply copied onto a plastic casting. Every microscopic three-dimensional surface detail is reproduced by the casting. It is a remarkably simple and dependable process that can be carried out in the field with very little training. (Ordinary forensic fingerprints are simply two-dimensional reproductions; plastic castings are three-dimensional copies of unique surface features.) Under an optical or electronic microscope, the casting appears as an irregular three-dimensional terrain.

Multiple plastic castings of the surface markings could be made, and these three-dimensional fingerprints can be shared between the inspecting and inspected parties. The major benefits of using plastic-surface castings are (1) they do not alter or otherwise harm the warhead or surface, (2) they are not left attached to the surface, and (3) they do not require intrusive inspector access to internal features of the warhead.

When the warhead is disassembled, the identifying marking of the casing can be re-fingerprinted and compared with the original plastic cast. Simply by keeping track of a unique identifying feature on the outer casing, assurance would be gained that a real nuclear warhead had been dismantled. Welded seals on storage canisters can also be authenticated by the three-dimensional plastic-casting technique.

Efforts by DOE national laboratories to counterfeit ("red-team") plastic-surface moldings were unsuccessful.

Other types of unique identifiers might be useful in authenticating treaty-controlled items. For instance, reflective-particle and electronic tags have been well developed, and they bring their own benefits for verification.

To maintain continuity of identification from beginning to end, the recommended procedure would be as follows: (1) A warhead removed from an operational missile could be "tagged" and/or "sealed" by an inspector with a unique identifier; (2) inspectors would check the assigned tags and seals at transfer and storage points; and (3) eventually, after a warhead is disassembled, the empty casing could be made available to inspectors, who would authenticate that the unique identifier for the empty container is the same as the original full-up warhead.

In order to ensure that sensitive information is not compromised, nations party to the agreement would have to agree on a common standard of visual access. In truth, it is unlikely that anything militarily useful would be divulged to other members of the nuclear club as a result of the inspection; nor are outsiders likely to gain proliferation-sensitive information directly or indirectly.

Tags and seals would be significant adjuncts to tracking warheads from point of deployment or storage. For warheads of uncertain provenance, more information — specifically their legitimacy as nuclear explosives — would be needed when undergoing disassembly verification.

Cloud Gap. In 1967 the Arms Control and Disarmament Agency, working with DOD and the AEC, initiated a field study code named Project Cloud Gap. Because much procedural and technical information about the study has been released, the report constitutes an excellent basis for evaluating technical verification of warhead dismantlement.

The project was intended to develop and test inspection procedures for monitoring destruction of nuclear weapons. Personnel in the project inspected 40 actual and 32 facsimile nuclear weapons, monitored the warhead destruction process, and assessed attempts to evade inspections. Although some minor classified information might have to be disclosed to inspectors, it was concluded that careful design and adherence to procedures would avoid disclosure risks for information that was truly sensitive.

Nuclear Materials

A comprehensive nuclear-reduction program is likely to require verification at each of several stages: special-materials inventory, means of production, locations of storage, and remnants from dismantlement. Much of the foundation for inspection and monitoring already exists in form of national and international means of safeguards. Additional material-control measures would be required to reinforce significant — deep — bilateral or multilateral reductions in weapons.

The pace and status of warhead disassembly will affect the means applied for verification. In recent years, Russia has reportedly been dismantling 2000 warheads per year, somewhat more than the United States.

In order to foreclose loopholes in verification of the process and results of dismantlement, inspectors would probably need to be present at intake and output facility portals during each dismantlement "campaign." Afterwards, by applying security seals, the containers for disassembled fissile components could be tracked to their destination for interim storage and ultimate disposition.

But, one would ask, what is inside an intact warhead or sealed container? To ensure that the original warheads or nuclear-material containers really (at least approximately) have the declared fissile quantities, several non-invasive means exist. Acceptable non-invasive means would be dependent on the level of secrecy maintained over warhead composition. At some point in the chain-of-custody it will be necessary to authenticate that the tagged object is actually a nuclear warhead or consists of essential declared components of a fissile core. In addition, it is important to ensure that untagged containers do not harbor forbidden or unaccounted nuclear components.

Other methods of analysis, such as precise measurement of emitted radiation, are available but less exact. In the case of weapons plutonium, it emanates detectable high-energy neutrons; however, accurate quantitative determination depends on geometric and material properties that are normally not divulged or available. Weapons uranium is even more difficult to measure and to quantify unless divided into small batches.

Technical-verification methods would have to be installed at production and treatment facilities in order to validate weapons-grade fissile materials fully declared and rendered according to treaty requirements. Stocks that are not full-up weapons are likely to be found in various conditions.

Acknowledged uncertainties have been a few percent in U.S. inventory, which translates into tons of weapons-grade materials. If that were the case for all superpower production, it would correspond to hundreds or even thousands of equivalent warheads. These uncertainties are a driving force for stringent and intrusive verification measures.

A Clinton Administration negotiated U.S./RF program for cooperation on nuclear security provides a model for installation of improved methods and instruments that would reinforce material protection, control, and accounting. The United States wanted to take a step ahead, towards mutual transparency and irreversibility. Proposed was an exchange of detailed information on aggregate stockpiles of nuclear warheads and fissile materials; all of this would be verified by reciprocal inspections — especially uranium and plutonium recovered from dismantled warheads.

Transparency in nuclear-materials monitoring has benefitted from appropriate technical demonstrations; for example, Kurchatov Institute and Argonne National Laboratory connected via the Internet to illustrate feasibility for remote monitoring of each other's stored nuclear materials. The U.S. DOE labs have been busy in the FSU, working with local scientists and technicians to improve safeguards at nuclear-material sites.

During the G.H. Bush administration, the United States offered to purchase 500 metric tons of enriched uranium from dismantled Soviet nuclear warheads, an offer consummated during the Clinton administration. As a result, American specialists were permitted to monitor the facility where Russia blended down the uranium enrichment so it can be used for commercial nuclear fuel.

Challenge inspections that supplement routine inspections are an essential tool to reduce uncertainties about production records. Short-notice challenges can be used to search for hidden stockpiles of warheads or fissile materials at locations not declared to be part of the official production complex.

Information sources external to the treaty, including intelligence collection, national technical monitoring and citizen reporting would also be of value in discovering locations for forbidden reserves.

A comprehensive verification program would aim to reconcile a multiplicity of production sources.

Inactivity at closed-down facilities would be easy to monitor. However, dual-purpose facilities that could produce weapon-grade materials would require verification or monitoring access. Complicating an arms-control regime is routine use of weapons-grade uranium for marine propulsion; various measures can be taken to develop confidence that the fuel is not diverted to weapons.

In any event, banned activities at facilities are relatively easy to detect — given intrusive access as warranted by treaty or agreement. In this regard, notice the retrospective trackability of centrifuge equipment that Iran and Libya had evidently purchased on the nuclear blackmarket. The centrifuge stages were contaminated with trace quantities revealing their original role, their final enrichment stage and their originator.

Nuclear-power plants and other peaceful nuclear activities need not be subject to additional restrictions for two major reasons: (1) because fissile materials used in peaceful activities, including power generation, are already or should be under extensive international safeguards, and (2) because these materials are not readily accessible or convertible in a manner that would actuate response to a design-basis security threat. Moreover, additional steps could be taken to demilitarize plutonium and uranium, making civilian sources of fissile materials even less suitable for utilization in weapons.

Secret Nuclear Testing

In order to determine if a nation has undertaken nuclear testing in secret, a number of unilateral and multilateral techniques exist to monitor compliance with existing treaties. National technical means were the foundation of nuclear-test detection and continue to be effectual in monitoring the current moratorium.

Nuclear testing is essential for gaining a major war-fighting arsenal of nuclear weapons. It is not necessary for maintaining existing nuclear armaments, nor was it patently required for Israel or South Africa in their secret development of small aircraft-deliverable stockpiles. Prohibition of nuclear-explosive testing serves to cap the capabilities of existing weapon-states and to curb potential rapid breakout from treaties that reduce or eliminate nuclear arsenals.

The Air Force Technical Applications Center (AFTAC) has operated the Atomic Energy Detection System (AEDS), a network of some 100 sites situated in more than thirty-five countries. AEDS was devised to detect nuclear detonations underground, underwater, in the atmosphere, and in space. AFTAC has monitored its AEDS sites twenty-four hours a day to assess compliance with the Partial Test Ban Treaty, Nuclear Non-Proliferation Treaty, Threshold Test Ban Treaty, and the Peaceful Nuclear Explosions Treaty.

As reported in the Brookings *Atomic Audit*, "AFTAC has operated in extreme secrecy [until 1997, when it] was given the lead role in collecting and assessing data to monitor compliance with the Comprehensive Test Ban Treaty. Under a CTBT, AFTAC would operate the U.S. National Data Center, which will feed information to and collect information from the new International Monitoring System, a proposed global network of 321 different sensors...." The global network will include 170 seismic stations (many now functioning), 80 automated radionuclide collection/assessment devices, 60 airborne infrasound detectors of acoustic low-frequency blast waves, and 11 hydroacoustic sensors in the ocean.

In addition, a ratified CTBT would allow on-site inspections of signatories through a challenge process based on national technical means or other evidence. A team of inspectors would go to or near the suspect site and might investigate for violations by making measurements of telltale radioactive materials and gases inevitably vented in nuclear explosions.

Upon entry into force, the CTBT will formally implement the International Monitoring System. A prototype network is already operating: Seismic stations of the International Monitoring System have been complete for several years in northern Europe, including the Russian arctic test site at Novaya Zemlya. These stations readily detected underground tests carried out by India and Pakistan in 1999, as well as two sets since conducted by North Korea.

An unprecedented joint statement by the American Geophysical Union and the Seismological Society of America has expressed confidence in the treaty's verification scheme.

In addition, an unofficial commission of scientists and experts collected suggestions on the verifiability of the test ban. The commission was established by the NGO VERTIC (see Volume 1) to "draw attention to the complex and constantly changing verification gauntlet that any potential violator would have to confront." Consisting of experts from 11 different countries and 14 scientists, selected for their expertise in one or more of the verification technologies or techniques requiring consideration, the commission's report "does not claim that a treaty is 100% verifiable, but it does make a strong case that there is a high probability that any event that might give rise to concern—as being a possible clandestine nuclear test—will be detected, located and identified." The commission expressed confidence that "explosions as low as 1 kiloton (and in some cases much lower) in all environments would be detected with a high degree of confidence...." They noted that the International Monitoring System, even though incomplete, "already has capabilities below 1 kiloton in some regions, particularly Central Eurasia."

VERTIC's commission highlighted the impressive capabilities of hydroacoustic and infrasound technologies, which are used to detect explosions, respectively, under the sea and in the atmosphere. These also help distinguish land-based underground explosions. Synergies of those two technologies with the worldwide seismic network and with radionuclide debris/gas monitoring reinforce the contention that nuclear explosions would be detected. If obtained during an on-site inspection initiated on the basis of other suspicious indications, radionuclide measurements would provide the "smoking gun."

The CTBT allows the international community to draw upon NTM to supplement the International Monitoring System. In addition, to help find evidence of a clandestine nuclear explosion, there are thousands of openly accessible scientific and environmental monitoring resources; these include increasingly available commercial satellite imagery and the global scientific seismic network. Improvements in the network are expected as the seismic stations go digital.

Evasion scenarios have been carefully evaluated. Decoupling a nuclear explosion or trying to hide it in another event (earthquake or construction) would be very difficult to accomplish without high risk of being discovered.

[**Nuclear Forensics Applied to Clandestine Weapons Testing**]. North Korea evidently tested a nuclear explosive in its mountainous northeast region on 9 Oct. 2006. Aside from the official North Korean announcement, various unofficial pronouncements indicated a relatively low explosive yield, less than a kiloton, suggestive of a malfunction.

Seismic data picked up at remote monitoring stations, corrected for regional wave-transmission anomalies, can be the basis for yield estimates.

Noble gases released in fission underground are difficult to keep from leaking; they can be detected in the atmosphere after being transported by wind outside North Korea. (Uranium and plutonium have different ratios of radioactive xenon isotopes, which can under good conditions provide data for confirmation of nuclear fission and for evaluation of fissile properties.)

Not in doubt, though, was the occurrence of a nuclear-explosive test (and another series in 2009) of the estimated yield at that time and place in North Korea.

Whether North Korea actually wanted the world to know of its capabilities is a question that deals with their internal political strategy.

The totality of verification resources available to the international community compounds the risks facing any potential treaty evader.

An interesting aspect of India's claim of having tested a "thermonuclear device" on 11 May 1999 is the public discussion that followed. A yield of 10-20 kilotons (later reduced to 10-15 kilotons) was estimated by U.S. seismologists. One suggestion is that India might have tested a "boosted primary" (rather than a two-stage thermonuclear explosive). India has declared that the test was for a two-stage thermonuclear weapon. What matters though, in the context of a comprehensive test ban, is that the nuclear explosions were not only detected by worldwide seismic networks, but also that estimates of the explosive yield were derived from public data.

Opponents to the CTBT fear that nations would be allowed to declare large territories off limits to approved inspections. In order to protect national security, the treaty authorizes up to 50 square kilometers to be designated as "restricted-access sites." However, "managed access" can be granted for inspectors to accomplish specific verification tasks within the chosen zone.

Missile and Bomber Constraints

Part of the response to a design-basis threat is an ability to assuredly deliver nuclear weapons to a target. At the same time, conflict-escalation beyond the immediate military or political goal is to be avoided. Aircraft, ships, and submarines are suitable platforms for a measured response to an attack, but land-based ballistic missiles are an exception to stability because of their short transit time and comparative vulnerability.

Thus, quantitative verification of limits — and destruction of strategic missiles, launchers, and ancillary facilities — would be fundamental to a treaty that embarked on deep cuts. High-confidence verification of sites, facilities, and treaty-limited items is relatively straightforward. Destruction is comparatively easier to monitor: Considerable practical experience was gained from the bilateral elimination of intermediate-range nuclear missiles under the INF Treaty.

One way of improving verifiability is by choosing configurations that are externally visible. For example, ballistic missiles reduced in size to carry only a single warhead are inherently suited for verification by measuring external characteristics, such as length and diameter.

Long-range cruise missiles, or bombers capable of launching them, would require more intrusive inspection, especially if the missiles were normally armed with conventional warheads. Warhead loadings on missiles and other delivery vehicles could be subject to agreed constraints; verification of RV count would require closer access, but would not call for excessive intrusiveness. Extensive studies and experiments showed that nuclear and conventional warheads could be externally distinguished in such a way that sensitive nuclear information would not be compromised, even while the cruise missiles were inside their launch tubes.

Unique identifiers could also be applied for reinforcing and expediting the tracking of warheads and delivery systems as they are moved from operational bases to dismantlement facilities.

For some "stand-down" agreements — such as de-alerting or de-targeting — specific actions, rather than objects must be validated; in that case, more complex and intrusive measures — if at all feasible — would be required to gain confidence in compliance. For instance, a stand-down from hair-trigger strategic retaliation favors operational configurations that are less reactive and more stable; these are conditions difficult to verify unless the warheads are separated from the missiles. Under such an arrangement, the warheads — with remote monitoring devices that confirm their detached status — could be stored at an agreed distance from launch sites.

Supplementary Verification

To guard against cheating or breakout as deep cuts are systematically undertaken, a stringent verification regime could be implemented. Experience to date indicates that militarily significant violations regarding nuclear-delivery systems are detectable on a timely basis, thus allowing various diplomatic and military measures to be invoked if necessary. However, verification of a total ban on nuclear weapons would require more rigorous endeavors, utilizing a broad repertoire of technical and procedural mechanisms.

Regardless of the means adopted, a transition period will be required to gain confidence in the accuracy of inventory declarations and the effectiveness of oversight. Nations could prepare for nuclear reductions by configuring their forces in such a way that transparency is enhanced.

In addition to the various formal means described earlier, especially for missile launchers, nuclear verification can be supplemented by enlisting outsiders. Here are two pathways.

Societal Verification. One measure advocated by (Nobel Laureate) Joseph Rotblat and others is termed "societal verification." It relies on citizens, especially scientists and engineers, imbued with moral obligation and reinforced with legal sanction, to report clandestine activities regarding weapons of mass destruction (indiscriminate casualty). This form of whistleblowing could be promoted and institutionalized if supportive international conventions and national statutes were draw up.

A principle flaw for the nuclear-weapons-free-world concept remains the difficulty of detecting attempts by a state or states to "break out" of their weapons-free status via clandestine efforts. The enormous destructive power of nuclear weapons makes it necessary to reduce the rescission probability to very near zero for undeclared weapons.

In order to address such concerns, Joseph Rotblat and others have advocated "societal" verification as an important complement to established means of technical and procedural verification. The basis for societal verification would be a socially responsible norm for community science and technology. As proposed by Rotblat, scientists and engineers would be asked to take an oath that a humanistic ethic will be considered in their technical work. Outlawed activities would be reported to a designated international organization. Because of improvements in Internet access and the ubiquity of portable phones, the ability to promptly, anonymously, and directly report violations is greatly enhanced.

In order to meaningfully attract whistleblowers, an international norm for ethical responsibility — supported by generous incentives and statutory immunity — would have to be instituted. Permanent international arrangements would be needed for protection, including opportunities for family asylum and the issuance of secure passports.

Experience to date on whistleblowing is mixed: One antithetical example is the terrifying experience of Israeli nuclear technician Mordecai Vanunu. Just the opposite is the case of South Africa's nuclear program: The underdeveloped morality of whistleblowing regrettably was not strong enough then to bring it to public light; instead, it was the government that freely acknowledged the program after-the-fact. On the other hand, increased scrutiny to nuclear enterprises in Iraq, Iran, Pakistan, and India might have resulted from information leaks that supplemented intelligence gathering.

Public-Satellite Images. Commercial imaging satellites are opening the potential for public verification of arms control. John Pike, at the Federation of American Scientists, was able with foundation funding to buy space-based photographs from the privately owned firm Ikonos. Other foreign-owned satellite companies are offering high-resolution pictures on the market. Photo images of otherwise-secret sites in Pakistan, India, Iraq, Iran, North Korea, Syria, China, the United States, and Russia have been made public.

Nations allied to the United States — France, Germany, Israel, Japan — are placing their own or dual-purpose, commercial and intelligence, satellites in orbit. As national, multinational, and international collaborations flourish for space-based reconnoitering, the potential is rapidly increasing for ubiquitous, nearly continuous monitoring of many anthropogenic activities. The wide public availability of refined GPS localization helps pinpoint locations of support facilities and activities.

These images and sensor data, with resolution adequate to discern small vehicles on the ground, can assist amateur as well as professional interpreters in confirming government claims about the purposes of facilities and quantities and types of armaments. While commercial depictions will not be as detailed as those available to governments from their own spy satellites, independent examination and evaluation will nudge government analysts and policymakers to be alert to impending risks in arms control and disarmament.

Iraq Invasion Postscript 5:
Verification Validity

Despite vigorous efforts by special forces assigned during and after the invasion, ***none*** of the alleged weapons or capabilities for mass destruction (indiscriminate casualty) were found in Iraq. A few chemical weapons that had been plowed underground after the Gulf War were uncovered from ammunition dumps.

To recap, the United States and Britain claimed they had intelligence information regarding a revitalized Iraqi nuclear-weapon program (including uranium and aluminum tubes allegedly for enrichment), stockpiles of usable and deliverable chemical and biological weapons, and a large number of banned missiles. None were uncovered during a thorough search by coalition forces.

[Fabricated Allegations of Iraqi WMD]. As detailed in Volume 2, Secretary of State Colin Powell's 2003 allegations before the UN were found to be baseless.

No such weapons were used against invading troops, nor were they found after occupation of Iraq. Powell had told the UN that Iraq was hiding the banned weapon-related biological, chemical, and nuclear systems.

The military-use allegations were unsubstantiated by UN inspections before the invasion or by coalition forces afterwards. Because of the incongruity of charges made, and the evident shifting of rationales, the charges appeared to have been fabricated in order to make the invasion palatable to the public.

No confirmation has emerged for any of the nuclear or chemical allegations that President Bush made in his 2003 State of the Union address. Despite supplying intelligence information to steer both the UN and post-occupation survey teams, WMD were not uncovered.

David Kay, a former UNSCOM chief inspector, was appointed to head up a U.S. post-invasion search team, but failed to find any WMD in the first half-year of search. Bush-administration officials attempted to make the best of the situation by selectively citing portions of the report that referred to "activities" and "equipment," none of which posed an actual threat or violated UN sanctions.

A conservative columnist, Dennis Byrne, thought that "Bush's critics need a reality check" because they are making "breathtaking distortions" in their "pronouncements about Kay's report." He asked readers to judge for themselves, charging that "any resemblance between what Kay actually said and much of what has been reported appears to be coincidental."

However, excerpts of the report shown in the interim-inspection report indicated that Bush's supporters took claims well out of context the claim that Iraq was in "tenacious pursuit" of WMD: No stocks of weapons were found. The best that the administration could demonstrate is what are described by Kay as "program activities" and "concealed equipment," which to Byrne and others prove "ambition" and "commitment," but not actual weapons.

Certainly the evidence of atrocities (ongoing for decades) by the Hussein regime has been blatant and confirmed. Not substantiated though, were allegations of Iraqi connections with Al Qaeda.

It is clearly evident that the pre-invasion UN and IAEA inspection teams were thorough and correct in their intrusive surveys of Iraqi facilities and properties. In fact, noting an absence of weapons of mass destruction and facilities for their creation, there could not have been a better real-life validation of the intrusive inspections that took place prior to the invasion.

Hans Blix, former head of UNMOVIC, has pointed out that since 1994, "not much [of WMD] was found and destroyed [in Iraq]." Nor did high-profile defectors lead to finding new weapons, although they did reveal — in particular — an undisclosed biological-weapons program that was aborted by the Iraqis.

Another conclusion of Blix: "We showed that it was possible to create an international inspection mechanism that was effective, that worked under the Security Council, and that was independent of intelligence agencies."

Of particular concern is the misleading intelligence information (in part from defectors) that was relied upon for the false allegations. It brings into question not only the value of such intelligence sources but also challenges the potential usefulness of societal verification because private motives can mislead analysts. Blix adds that the intelligence from satellites and defectors "did not lead us to interesting [weapon] sites."

As presented in Volume 1, a long and well-documented history of inadequate and distorted intelligence collection and analysis exists from the Cold War. Various claims about Soviet capabilities proved to be overblown or highly distorted. Sometimes high-level officials intervened to cast a protagonist's viewpoint. This was particularly blatant with the actions of the U.S. government-appointed "Team B" that began in 1976 to distort intelligence findings. A cascade of worst-case revisionist interpretations followed, alleging that the Soviet threat was underestimated.

Team B (with brash members Richard Cheney and Paul Wolfowitz) and the Committee on the Present Danger, propagated many inaccuracies and distortions, such as a magnified number of Soviet Backfire bombers, false evidence for a Soviet anti-acoustic submarine, alleged Soviet encouragement of revolutionary violence and terrorism, the abbreviated time it would take for a "rogue" nation to develop an ICBM, false allegations about China's military spending, and — more recently — the hyperbolic estimate of Iraq military potency.

[Interim Inspection Report (2003)]. (Excerpts from public statement)

We [the Iraq Survey Group, ISG, led by David Kay] have not yet found stocks of weapons, but we are not yet at the point where we can say definitively either that such weapon stocks do not exist or that they existed before the war and our only task is to find where they have gone....

We have discovered dozens of WMD-related program activities and significant amounts of equipment that Iraq concealed from the United Nations during the inspections that began in late 2002....

With regard to **biological warfare** activities, which have been one of our two initial areas of focus, [Iraq Survey Group] teams are uncovering significant information — including research and development of BW-applicable organisms, the involvement of Iraqi Intelligence Service (IIS) in possible BW activities, and deliberate concealment activities....

We are starting to survey parts of Iraq's chemical industry to determine if suitable equipment and bulk chemicals were available for chemical weapons production....

Iraqi practice was not to mark much of their chemical ordinance and to store it at the same [Ammunition Storage Points] that held conventional rounds, the size of the required search effort is enormous....

Multiple sources with varied access and reliability have told ISG that Iraq did not have a large, ongoing, centrally controlled CW program after 1991. Information found to date suggests that Iraq's large-scale capability to develop, produce, and fill new CW munitions was reduced — if not entirely destroyed — during Operations Desert Storm and Desert Fox, 13 years of UN sanctions and UN inspections....

Saddam Hussain remained firmly committed to acquiring nuclear weapons. These officials assert that Saddam would have resumed nuclear weapons development at some future point. Some indicated a resumption after Iraq was free of sanctions....

Despite evidence of Saddam's continued ambition to acquire nuclear weapons, to date we have not uncovered evidence that Iraq undertook significant post-1998 steps to actually build nuclear weapons or produce fissile material....

With regard to **delivery systems**, the ISG team has discovered sufficient evidence to date to conclude that the Iraqi regime was committed to delivery system improvements that would have, if OIF [Operation Iraqi Freedom] had not occurred, dramatically breached UN restrictions placed on Iraq after the 1990-91 Gulf War....

In any event, it appears that the United States and Britain, in striving to justify the invasion of Iraq, did not adhere to rigorous standards of intelligence corroboration.

The Arms Control Association called for a broadening of the non-proliferation "campaign," making five suggestions for additional steps to be considered.

▶ First, "creation of a long-term, multinational fuel supply that would make national possession of uranium-enrichment plants unneeded and uneconomical."
▶ Second, revising lax export-control systems [all of them, not just Pakistan's].
▶ Third, "quash" DOE's "nuclear research" programs that "actually promote the spread of reprocessing technology and the means to produce plutonium" [a suggestion with which we demur, because the research is needed to help destroy weapons plutonium].

▸ Fourth, U.S. reaffirmation of support for a fisban.

▸ Fifth, "meaningful limits on U.S. nuclear weapons capabilities [e.g., earth penetrating nuclear weapons]," and support for the CTBT and the verifiable dismantlement of "excessive U.S. and Russian nuclear bombs and missiles."

In short, the Arms Control Association advised that

> a more comprehensive and robust non-proliferation strategy ... requires more than just pressure on a few nuclear "have-nots" — it requires greater restraint and leadership from the nuclear "haves."

As for the future, Blix implied that verification, monitoring, and inspection processes have been bolstered by the decade-spanning Iraq intrusion, and the experience will be valuable for future disarmament.

Regrettably, the resources used in seeking WMD could have been applied with equal zeal to ferreting out hidden stockpiles of Iraqi conventional weapons, ammunition, and explosives with a considerable savings in injuries and lives of soldiers and noncombatants during the subsequent post-invasion insurgency. The insurgents took advantage of the Bush-administration fixation on alleged WMD.

The boxes on the following pages follow reflect independent findings about the validity of pre-invasion verification. The inspection regime was thorough, vigilant, incisive, and accurate.

[The Last Word on Iraq's Alleged WMD]. Charles Duelfer, the CIA's special adviser to the Iraq Survey Group (ISG), told the Senate Armed Services Committee Oct. 6 [2004] that Iraq destroyed its stockpiles of chemical and biological weapons, as well as eliminated its nuclear weapons program, after the 1991 Persian Gulf War. Although his findings thus far largely confirm previous reports, they offer the most extensive analysis to date of the state of Iraq's weapons of mass destruction (WMD) before last year's U.S.-led invasion....

The testimony and report show that deposed Iraqi leader Saddam Hussein, constrained by UN sanctions, had not restarted the country's nuclear, chemical, or biological weapons programs. However, he was seeking to preserve and restore, to varying degrees, the intellectual and physical capacity to resume the nuclear and chemical weapons programs if sanctions put in place by the UN Security Council after the Gulf War were lifted.

Escaping the sanctions was a "top priority" for Hussein, who manipulated the UN oil-for-food program by granting rights to low-priced Iraqi oil in exchange for recipient countries' support for getting the sanctions lifted. Established in 1995, the program allowed Iraq to purchase food, medicine, health supplies, and other civilian goods with proceeds derived from oil sales, which were held in a UN escrow account.

Iraq had some success in circumventing the sanctions, the report said. Baghdad was able to obtain cash through illegal oil sales and the import of illicit goods, including some dual-use items useful in WMD programs....

In any case, the sanctions were largely effective at restraining Iraq's weapons programs. Duelfer told the committee that the sanctions both constrained Iraq's weapons-related imports and induced Hussein not to pursue WMD because such efforts would jeopardize his goal of getting the sanctions lifted in the past, the report says. UN inspectors had cooperated closely with U.S. intelligence during the 1990s. Secretary of State Colin Powell told the Security Council in early February 2003 that Iraq attempted tap the inspectors' "communications."

The report helps explain some pre-war intelligence failures, revealing that Iraq often moved conventional military assets to hide them from potential attacks, but U.S. intelligence often assumed Iraq was actually moving illicit weapons.

[Alleged Iraqi WMD].

Nuclear. [The] ISG "found no evidence to suggest concerted efforts to restart the [nuclear] program" after 1991. Although Duelfer testified that "Saddam did not abandon his nuclear ambitions," he said that Iraq's ability to produce nuclear weapons and retain the relevant personnel were in "decay" as a result of the sanctions.

Duelfer also definitively refuted two elements of the Bush administration's pre-war case that Iraq had reconstituted its nuclear weapons program. First, the ISG found no evidence that Iraq tried to procure uranium from other countries, instead learning that Iraq refused a private offer to help it obtain uranium from the Congo. Second, Duelfer testified that 81mm aluminum tubes Iraq was trying to import were solely for conventional rockets. The administration had argued that the tubes were likely to be used in centrifuges for enriching uranium. A July Senate Intelligence Committee report found the intelligence underlying both the uranium and tubes claims to be weak.

Chemical. Iraq destroyed its chemical weapons stockpiles in 1991, according to the report, but Hussein still intended "to resume a [chemical weapons] effort when sanctions were lifted." Iraq was increasing its chemical production infrastructure, which provided "the inherent capability" to produce chemical weapons in the future, Duelfer said, adding that Iraq would have been able to produce mustard agent within "months" and nerve agent "in less than a year or two." However, the ISG has not "come across explicit guidance from Saddam on this point," Duelfer said, and there is no evidence that Iraq attempted to procure "precursor chemicals in bulk," according to the report....

Biological. According to the report, Iraq "appears to have destroyed" its biological weapons and bulk weapons agent in 1991 and 1992. Although Baghdad maintained a research and development program until 1996, it then abandoned it. Since then, there appeared to be "a complete absence of discussion or even interest in [biological weapons] at the presidential level," the report says....

Duelfer told the committee that the ISG has found no evidence that Iraq possessed mobile biological agent production facilities — another claim the administration had advanced prior to the war. Expressing perhaps the most definitive judgment to date on two trailers discovered shortly after the invasion, Duelfer stated that "they have absolutely nothing to do with any biological weapons." A May 2003 CIA report judged that the trailers were likely for weapons agent production....

The report's findings are consistent with those of the UN inspectors, who returned to Iraq in November 2002 and remained until the invasion. The report does, however, describe WMD-related activities that the Iraqis failed to declare to the United Nations, such as several missile research programs and attempts to procure related materials.

[The Weapons]. Duelfer testified that Iraqi WMD "stocks do not exist," despite occasional finds of pre-1991 chemical munitions. He also said that the ISG has found no evidence that WMD were transferred to other countries, a theory some administration officials have advanced....

Despite the ISG's findings, President George W. Bush ... continued to defend the invasion. He stated Oct. 7 [2004] that Duelfer's report proved that Hussein "retained the knowledge, the materials, the means and the intent to produce weapons of mass destruction."

Bush administration officials, however, claimed numerous times before the invasion that Iraq actually possessed illicit weapons. Moreover, Duelfer's testimony and report are more nuanced than Bush's statement suggests. For example, Hussein appeared to have intentions to develop certain types of weapons but lacked the capabilities. In other cases, the Iraqi leader had some residual weapons capabilities but no evident intentions to make use of them.

The report's findings are consistent with those of the UN inspectors, who returned to Iraq in November 2002 and remained until the invasion. The report does, however, describe WMD-related activities that the Iraqis failed to declare to the United Nations, such as several missile research programs and attempts to procure related materials.

****

Thorough, formal verification of the dismantlement of nuclear weapons would be essential to obtain confidence about nuclear disarmament. The process would have to start with operational weapons, tracking them to through stages of removal from service, transport to holding sites, and into storage pending disassembly. When moved to dismantlement sites, the chain-of-custody would have to be verified, as would be ultimate disassembly into subunits and constituents.

Existing stockpiles of fissile materials, as well as that surplused from earlier dismantlement, would likewise have to be monitored and checked. Treaty-legitimated verification would encompass other aspects of potential nuclear arsenals, including new developments in testing rockets and explosives.

Although this Verification Issues Section has concentrated on how to provide an essential level of confidence for nuclear arms-control commitments — especially deep cuts — some symbiotic benefits might accrue from parallel verification regimes. Examples include verification measures for prohibited conventional and chemical weapons. As in the case of UNSCOM in Iraq, inspection teams could be empowered to make concurrent observations about other activities or equipment germane to relevant arms-control or disarmament jurisdictions.

The negative findings of UNSCOM in Iraq were faulted and criticized vehemently by hawks in and out of the American and British governments, who desperately sought justification for their destined invasion of Iraq. The highly politicized invasion proceeded without explicit UN authorization or solid evidence of WMD transgressions. Post-invasion access and intensive scrutiny failed to validate any of the claims put forth. The pre-invasion verification regime imposed upon Iraq has proven to be a prime example of the thoroughness and validity of intrusive challenge inspections.

Aside from the cautionary fact that intel and verification reports were ignored and politicized, the sequence of Iraq inspections and their supportive infrastructure proved themselves to be highly effective in evaluating armament details and military significance. The post-invasion follow-up validated pre-invasion assessments.

Had intrusive verification been negotiated during the Cold War for either the CWC or BWC, the massive highly secret biological weapons program of the Soviets would probably have been detected. This creates a classic conundrum: It has been difficult in the past to negotiate formal verification agreements because, from the viewpoint of someone who has something to hide, verification can be too effective!

To augment treaty-authorized national and international means of verification, the public at large can be enlisted to help (just as defectors assisted Cold War intelligence collection). Already modern technology and globalization are unintendedly supplementing formal verification measures and unofficial openness trends.

G. NEGOTIATED NUCLEAR DISARMAMENT

However one might interpret its implementation, mutual deterrence was probably instrumental in the evaporative—rather than explosive—finish to the Cold War. Even so, the same deflation in confrontation might have been peacefully reached with much lower levels of armaments, accompanied by much less mutual risk and lower budgetary stress.

The international-security environment has since stabilized enough that further improvements could be achieved, with all countries maintaining national security. Recognizing that at least one of the former ideological and military adversaries (Russia) is ready, anxious and hungry for significantly reduced armaments, we must ask rhetorically what strides might be undertaken? In particular, what are the measured steps that are verifiable— satisfying the universal notion that processes for improving mutual security should be observable?

Recall that a vector is a mathematical symbol depicting a course of action. A vector can embody not only direction, but a meaningful magnitude, namely the speed intended in reaching the goal.

In the minds of many advocates, the nuclear vector points toward disarmament. Perhaps that's not rapidly approachable nor ever achievable, but — for managing weapons of indiscriminate casualty — disarmament is a legitimate goal, to be sought with forbearance via negotiated measures. The Non-Proliferation Treaty's Article VI, which was reaffirmed at the NPT 1995 Review Conference, set the foundation for universal nuclear disarmament. In contrast to the abrupt "quantum" jump sought by impatient abolitionists, we — the authors of this book — believe and assume that it's more feasible to retain a semblance of quasi-static equilibrium: Each stage of reductions, as the participants advance (or even jump) along the inclined vector, should proceed one or a few steps at a time.

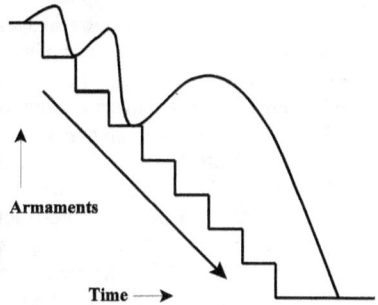

One of the most reasoned analysts was George Kennan, who — reflecting on 40 years of containment — warned in 1987 that

> ... I see the weapons race in which we and they are now involved as a serious threat in its own right, not because of aggressive intentions on either side but because of the compulsions, the suspicions, the anxieties such a competition engenders, and because of the very serious dangers it carries with it of unintended complications — by error, by computer failure, by misread signals, or by mischief deliberately perpetrated by third parties.
>
> For all these reasons, there is now indeed a military aspect to the problem of containment as there was not in 1946; but what most needs to be contained, as I see it, is not so much the Soviet Union as the weapons race itself. And this danger does not even arise primarily from political causes. One must remember that while there are indeed serious political disagreements between the two countries, there is no political issue outstanding between them which could conceivably be worth a Soviet-American war or which could be solved, for that matter, by any great military conflict of that nature.

Numerous proposals for controlling and reducing nuclear arms have been advanced, many of them mentioned earlier in this book. This Chapter sorts through them, outlining the most credible, emphasizing those that would enhance negotiated time-bound nuclear disarmament.

First, a caveat: nuclear disarmament does not equate to total disarmament. The management and control of conventional weaponry are outside the purview of this book. Nations will retain whatever level of conventional forces and armaments that they deem necessary to withstand outside interference or to carry out their political aims.

A comprehensive arms-control action plan for the 21st century was formulated in a July 1999 report of the quasi-official Tokyo Forum for Nuclear Nonproliferation and Disarmament, an independent panel consisting of disarmament experts, government officials, diplomats, and military strategists from around the world.

[Key Proposals of 1999 Tokyo Forum].
- ▸ Reaffirm the central bargain of the Nuclear Non-Proliferation Treaty
- ▸ Eliminate nuclear weapons through phased reductions
- ▸ Bring the nuclear test ban into force
- ▸ Revitalize START and expand the scope of nuclear reductions
- ▸ Adopt nuclear transparency measures
- ▸ Remove nuclear weapons from hair-trigger alert
- ▸ Control fissile material, especially in Russia.
- ▸ Prevent weapons of mass destruction from falling into hands of extremist groups
- ▸ Strengthen measures against missile proliferation
- ▸ Exercise caution on missile defense deployments
- ▸ Stop and reverse proliferation in South Asia
- ▸ Eliminate weapons of mass destruction in the Middle East
- ▸ Reduce nuclear and missile dangers on the Korean Peninsula
- ▸ Exercise Security Council vetoes to support nonproliferation
- ▸ Revitalize the UN Conference on Disarmament
- ▸ Strengthen verification for disarmament
- ▸ Create effective compliance mechanisms for nuclear nonproliferation/disarmament

The Tokyo Forum made seventeen specific recommendations to reduce nuclear dangers and to stop and reverse nuclear proliferation. Their 17 key proposals are summarized in the list below. Regarding control of fissile materials, the Forum suggested the specific items listed in the box that follows.

[Fissile-Material Controls].
- ▸ an end to production of materials for weapons purposes
- ▸ expediting negotiation of a fissile-material cut-off treaty
- ▸ increasing transparency of fissile stockpiles
- ▸ preventing terrorists from acquiring dangerous materials
- ▸ improving material protection and control
- ▸ strengthening threat-reduction programs in Russia
- ▸ expanding fissile material verification and safeguards

Nuclear-weapon states are beginning to give at least lip service to drastic cutbacks. The 1998 Labour government of the UK had a "New Agenda" which called for "speedy, final and total elimination of nuclear weapons."

In the mid-1990s, the United Nations assembled a spectrum of experienced specialists who suggested ways to move toward "real" nuclear disarmament. Taking into account that the nuclear arms race had come to an end, they counseled measures to reduce the importance of ballistic missiles and to facilitate discussion of "serious" nuclear disarmament. The experts affirmed that the road to reductions should not be unilateral or asymmetrical, and it should include the specific items compiled in the next box.

[UN Experts' Advice].

Unconditional pledges not to threaten or use nuclear weapons against any party:
▸ setting up missile-free zones
▸ organizing a public campaign to stigmatize ballistic missiles

Engagement in dismantlement talks:
▸ negotiating START III
▸ taking more strategic weapons off of alert status
▸ declaring inventories of weapons and materials
▸ extending the process to other nuclear-weapons states

The Stimson Center, an NGO "community of analysts devoted to offering practical solutions to problems of national and international security," assembled in 1999 a Committee on Nuclear Policy that produced a report called "Jump-START." The Committee was formed from nuclear-weapon project directors of NGOs in the United States and Europe, as well as retired military leaders and national lawmakers.

Their recommendations relate to implementation of deep reductions, removal of hair-trigger ballistic-missile response, and improvement of fissile material and warhead control. Because of "serious concerns" about the safety, control, and overabundance of nuclear weapons, the Committee called for the United States and Russia to engage in a number of arms-control and non-proliferation endeavors.

Jonathan Dean, an experienced diplomat and arms-controller, emphasized that disarmament must be definitive, long-term, and irreversible. The measures he considered important are condensed in the list that follows.

[Dean's List].

Urgent measures to reduce current risks:

▸ remove all warheads from operational deployment
  (and transfer them to internationally monitored storage)
▸ establish reciprocal monitoring of stockpiles and fissile materials
  (in order to reduce the risk of forcible seizure, theft, or illegal sale)
▸ destroy all missiles withdrawn from operational deployment

To implement these recommendations, Dean foresaw the need for:

▸ nuclear-weapons states to act together
▸ the NPT be improved and become universal
▸ weapons and delivery systems to be neutralized
  (by placing them into monitored storage)
▸ UN and regional institutions to be radically improved

[Japanese Government Global Nuclear Disarmament Benchmarks (2009)].

A 2009 update of Japanese government support for moving towards zero nuclear weapons contains 11 "benchmarks" based on three major "pillars": nuclear disarmament, international non-proliferation, and peaceful uses of nuclear energy.

The first pillar, nuclear disarmament by all nuclear-weapons-holding states, contains five benchmarks:

• Leadership and cooperation between the United States and Russia, including a successor treaty to START I at an early date, further reductions in nuclear warheads, mutual confidence regarding missile defense, and strengthening the framework for controlling nuclear weapons and material.

• Nuclear disarmament by China and other nuclear-weapons-holding states.

• Transparency over nuclear arsenals, including regular and sufficient information numbers of weapons, fissile material, and delivery vehicles.

• Irreversible nuclear disarmament, including dismantlement of nuclear warheads, nuclear-testing sites, and production facilities.

• Studying verification because highly accurate verification of weapons dismantlement will be required.

The second pillar contains three multilateral measures:

• Banning of nuclear tests and ratification of the Comprehensive Nuclear Test Ban by the United States, China, India, and Pakistan.

• Forbidding production of fissile material for nuclear-weapons purposes, starting with resumption of negotiations on the Fissile-Material Cut-off Treaty with an interim moratorium.

• Restrictions on ballistic missiles capable of delivering a nuclear warhead.

The third pillar with three benchmarks is based on peaceful uses of nuclear energy:

• International cooperation in civil nuclear energy, including safeguards, safety, security, and fuel-supply assurance with IAEA regulation.

• Enhanced and comprehensive IAEA safeguards.

• Prevention of nuclear terrorism, which implies strengthening international control of nuclear materials.

Assuming disarmament makes headway, all aspects of nuclear arsenals and their delivery systems would require attention. For example, the United States has been tagged as "the world's main proliferant in air-breathing cruise as well as long-range missiles." One reason was that, despite termination of Cold War hostilities, the United States continued "to adhere to a strategy of nuclear deterrence based on large-scale deployment and possible use of long-range ballistic missiles." Also the United States has employed nuclear-capable cruise missiles armed with conventional high-explosive warheads for selective targeting (as in Iraq, Sudan, Yugoslavia, and Afghanistan). Thus, not only designated strategic weapons and delivery systems, but also dual-purpose systems need to be dealt with in the context of force reductions.

As a nuclear specialist, I can highlight more than half-dozen generic steps — major categories of disarmament — each suitable for gradual implementation. If these recommendations were followed, considerable headway would be made with the inherited arms-control agenda, specifically including a formalized ban on nuclear testing and reductions in strategic arsenals (see distilled list below).

Recommended Short-Term Focus and Voluntary Priorities

▶Constraining missile-launches and restricting long-range missiles
▶Limiting the number of warheads loaded on ballistic missiles
▶Enhancing cooperative missile defense
▶Keeping outer space free of weapons
▶Transitioning to low inventories of warheads
▶Disposing of nuclear-weapon materials
▶Converting and reducing the defense infrastructure

Another way of parsing the policy recommendations is to divide them into nine discrete goals under the overarching theme of nuclear de-emphasis, as given in the box below, making use of the Latin prefix *de-* (meaning reversal or removal), and taking slight liberty with the English language.

> **[Nuclear "De-Emphasis"].** Nine specific steps or goals can be identified to reverse (or de-emphasize) the Cold War instruments of nuclear destruction:
> 1. "de-targeting" nuclear aim-points
> 2. "de-alerting" missile-launch systems
> 3. "de-mating" nuclear warhead s from delivery systems
> 4. "de-MIRVing" ballistic-missile reentry vehicles
> 5. "de-creasing" nuclear arsenals
> 6. "de-fending" against ballistic missiles
> 7. "de-weaponizing" outer space
> 8. "de-militarizing" fissile materials
> 9. "de-minishing" the inherited nuclear infrastructure

Here's one way to make headway en-route to disarmament: Start by making constructive reaffirmation or partial revision of obligations embodied in NPT Article VI, the long-standing agreement to pursue "general and complete disarmament," including the prohibition of nuclear weapons. Accompanying such a reinforced declaration could, for example, be a demonstrable reduction in convulsive launch readiness by de-targeting and de-alerting. These are the first significant steps that constitute the beginning of nuclear de-emphasis. The next logical measure would be a change in operational environment such that warheads and missiles are separated (de-mated) — preferably relocated at a significant distance from each other.

The remainder of this Section is devoted to filling in details and explaining the recommended goals for nuclear de-emphasis. Some are linked together, and several can be achieved somewhat concurrently.

Missile-Launch Constraints

Maintaining nuclear forces at a low alert level is not a new operational practice. In the late 1940s and 1950s, the major strategic nuclear weapons of U.S. forces were kept in storage bunkers; bombers went on routine training exercises without payloads and were not airborne for extended flights. To launch a nuclear strike would have required hours or days of preparation. Also, nuclear munitions were always stored away from the aircraft, and the earliest U.S. bombs had their fissile material separated from the rest of the warhead as an added safety measure. A provision of the START I Treaty prohibits storage of nuclear weapons within 100 kilometers of a conventional-bomber base.

High-alert situations, in which nuclear weapons are launchable within a few minutes, started at the end of the 1950s, when some bombers — fueled and armed — were maintained in constant readiness at the edges of runways. After the Berlin Crisis in 1961, the United States kept a portion of its bomber force constantly in the air, loaded with nuclear weapons; in the early 1960s, thousands of nuclear-armed missiles and aircraft were placed on high alert.

Other nuclear-armed nations, concerned about the survivability of their deterrent forces, formulated similar policies and practices. The Soviet Union is believed to have had periods of heightened defensive readiness. The United Kingdom's 15-minute bomber-deployment plan was called Quick Action Alert. France also had a similar ground-alert capability (bombers ready to take off with 15-minute warning).

By 2001 — the end of the Clinton administration — tensions from the past had unwound, and it had become possible for nuclear-weapons states to engage in systematic steps toward strategic stabilization. Accompanying the improved mood was a growing recognition that strategic deterrence had become relatively stable and reliable. The actual and perceived likelihood of surprise nuclear attack was greatly diminished.

This harmonious environment had already led to voluntary constraints in missile-launch policy, such as declarations that population centers were removed as aim-points ("de-targeting") and that strategic delivery forces had been put at reduced levels of alert ("de-alerting"). These are the first two of nine steps recommended in this book as part of a comprehensive program for weaning weapon states off their nuclear dependency.

To solidify these stabilized conditions, visible and verifiable constraints (launch obstacles) would have to be installed at missile structures. Short of stopping production or destroying weapons, the next measure— whether unilateral or negotiated — would be the physical separation ("de-mating") of warheads from their delivery systems (the third recommended step).

Unfortunately, the atmosphere changed dramatically in 2002: To Russia's dismay, the G. W. Bush administration withdrew the United States from the ABM treaty and announced its determination to develop and deploy a missile-defense system. Russian leaders had been hoping for an opportunity to divert resources from defense activities to bolstering their civilian economy, but the ABM treaty demise greatly strengthened the hand of militarists: Weapons being decommissioned were stored instead of destroyed, and delivery-system development and procurement increased.

With great fanfare in 2002, Presidents Bush and Putin signed the Strategic Offenses Reduction Treaty, known as the Moscow Treaty. Under that agreement, each side was to reduce its "operationally deployed" strategic missiles gradually, from about 6000 down to around 2000—which sounds impressive until you look at the fine print. It turns out that the missiles were not to be destroyed, but merely disassembled and put into storage. These changes, to be accomplished by 2012 when the treaty expired, would leave the parties free to reactivate their arsenals. The parties could withdraw earlier—as the United States did from the ABM accord.

The Moscow Treaty had no specific verification provisions, even though verification is essential to foster confidence in strategic stability.

Alexei Arbatov, a contributor to Russian strategic policy thinking, proposed in 1999 a program of gradually improving "transparency" by assigning U.S. and Russian inspectors to each others' bases and command-and-control centers. This, Arbatov believed, would "make surprise attack even physically impossible and remove the need for launch-on-warning [and] launch-under-attack ... options and exercises."

Physical constraints on missile launches have two primary objectives: to reduce the likelihood of accidental or unauthorized use, and to provide time for clarification, reconsideration, or negotiation. Properly implemented, such measures maintain strategic stability and reinforce deterrence. If necessary, an orderly return to alert status could be made without a decrease in safety, security, or reliability of the weapons. Above all, the measures must be practical and effective, and it should be relatively easy to clearly verify that the weapons remain in conformity with agreed constraints.

Crisis stability exists when there is little or no incentive or ability to launch a hasty nuclear attack if political relationships deteriorate rapidly. Stability is enhanced if most nuclear forces have been deactivated so that lengthy procedures, such as warhead re-mating, would be required to make the weapons operational. For crisis stability, it is necessary to ensure that deactivated warheads are not vulnerable to a pre-emptive attack—for instance, all not stored in one place. Proposals in *The Nuclear Turning Point* are designed to accommodate such concerns. Active missile forces would be deployed in a way that would not leave them vulnerable to a plausible surprise attack, and inactive forces would be dispersed and incapable of being launched on short notice. Various alternative options could be exercised, as designated by treaty participants.

Nuclear Jeopardy. Quick-response nuclear postures remain in effect: Thousands of nuclear missiles are prepared for rapid re-targeting. Both the United States and Russia have roughly 3000 strategic nuclear warheads poised. Moreover, launch-on-warning strategy and doctrine continue.

Russia's current difficulties in maintaining and deploying its most survivable forces—submarines at sea and mobile rockets on land—intensify reliance on a hair-trigger posture. It requires humans to make momentous decisions under severe time constraints, with only minutes available from the warning stage to a decision point that risks catastrophe.

Exacerbating the situation is strategic reliance on missiles with multiple warheads, which themselves are inherently high-value targets. To protect the missiles, military commanders keep them on high alert, ready to launch within minutes of warning.

The higher the military-alert level, the greater is the risk of unauthorized launches. If the alert level escalates as weapons are primed for rapid launch, fewer procedures are needed to implement an attack, either authorized or unauthorized, deliberate or mistaken. An NGO group assembled by the Union of Concerned Scientists characterized the situation as follows:

> With nuclear missiles armed, fueled, and ready to fire upon receipt of a few short computer commands, the need for strict safeguards to prevent unauthorized launch is obvious. But no safeguards are foolproof, and maintaining nuclear forces in a way that required additional physical steps to launch a missile would offer greater protection.

Ironically, having a large number of missiles on a low operational-alert is more stable than having a small number of missiles on high alert. Alert status can be further reduced by forsaking automated-response options that include launch-on-warning and launch-under-attack.

Jozef Goldblat, a (French) consultant to the United Nations Institute for Disarmament Research (UNIDIR), preferred additional measures:

> All strategic forces should now be taken off alert. Such an undertaking could be followed by an observable separation of nuclear warheads from delivery vehicles in such a way as to render their use physically impossible without a considerable delay facilitating detection of preparations for use.

In addition to the ever-present danger that a crisis might escalate into nuclear conflict, there is a risk that missiles might be launched by error, by accident, or without authorization. That jeopardy has motivated a series of policy decisions, executive agreements, and formal treaties between the United States and the Soviet Union.

Noteworthy strategic-stabilization measures recommended in 1997 by Sam Nunn (former Senator) and Bruce Blair (heading the Center for Defense Information) had an expectation of reciprocity. Launch constraints, some self-imposed and some verifiable, serve to reduce the spontaneity and reflexiveness of nuclear-weapons delivery.

Confidence building, in the context of launch restraints, consists of (1) meaningful, but easily reversible measures, such as establishing communication links — hot lines — between parties, and (2) agreements that require notification of potentially provocative events, such as test launches of ballistic missiles.

Command and control "reinforcement," singled out by Nunn and Blair, consisted mostly of one-sided actions taken to keep the nuclear button tightly linked to the reins of the civilian-military chain of command. These safeguards maintain the security of administrative and technical launch procedures, typically enabled by a special key or code and the simultaneous insertion of two or more keys or codes. Reinforcement could be achieved by transferring custody of the keys or codes from launch personnel to a separate organization. Additional assurance would be provided by installing capability for post-launch in-flight self-destruct and by missile "de-targeting" (impact-change).

Pre-launch safety devices that initiate an alarm or terminate launches when the procedure is illegally initiated would enhance confidence (but be difficult to verify). A post-launch safety feature could include a self-destruct mechanism installed on a missile; if provision were made for outsiders to openly monitor (but not defeat) destruct-signal communication channels, post-launch monitoring would be a useful means of ensuring that an errant missile would be destroyed or diverted from its programmed target.

Constraints on aircraft are not discussed here. Because controls over the arming, takeoff, flight, and recall of piloted bombers are much easier to introduce, they do not undermine stable deterrence.

De-Targeting. For the purposes of this discussion, "de-targeting" means "changing, the programmed impact coordinates (aim-points) of long-range ballistic missiles." Under U.S. SIOPs, the missiles were normally aimed at so-called strategic targets, likely to be military installations — but population centers were not necessarily excluded. Changing programmed impact coordinates to some point in the ocean reduces the risk that an unauthorized or accidentally launched missile would reach a populated area.

Presidents Clinton and Yeltsin in 1994 made a symbolic pledge not to aim strategic missiles at each other's country, announcing that Russian and U.S. missiles were no longer "aimed at children." But the missiles, according to Bruce Blair, retained their wartime targets in computer memory, "which can be activated in seconds by a few computer strokes." Thus the change, though welcome, had little lasting military significance.

Russians derisively referred to the Yeltsin-Clinton re-targeting agreement as a "zero flight plan."

Blair contends that President Clinton "seriously misrepresented the effect of his so-called de-targeting pact with then-President Yeltsin." Blair estimated that about 2000 strategic warheads on each side remained on hair-trigger alert, and he advocated that these Cold War holdover practices be ended — to reduce tensions, minimize risks of mistaken or unauthorized launch, and strengthen nonproliferation.

De-Alerting. Missile-launch "de-alerting" would be one of the easiest high-payoff ways to reduce nuclear dangers. De-alerting is "the use of procedures or reversible physical constraints that increase the time or effort to launch a strategic ballistic missile." This goes well beyond de-targeting.

De-alerting land-based missiles can be reinforced by ensuring that the keys for missile launch are securely locked away from immediate or unauthorized access. While attractive as a means of strategic stabilization for circumspect adversaries, the inherent reversibility of de-alerting is the main factor limiting its usefulness. Although it gives a warm, fuzzy feeling, it has few observable features. While necessary for strategic stability, missile-launch de-alerting is not sufficient by itself.

Jonathan Dean supported de-alerting as a further step in missile control. Defining de-alerting broadly as "agreed measures to delay launch of nuclear-armed missiles," he characterized the policy as a "potentially valuable way of reducing the continuing dangers arising from ... missiles, dangers that include unauthorized launch, launch by error, or accidental launch culminating in large-scale launch-on-warning." Dean advocated that de-alerting be extended step-by-step to other nuclear-armed nations, and that missile-free zones be established. He also believed that it would be valuable to have "worldwide missile warning systems giving information on missile launches, including tests...."

Bruce Blair posited that crisis stability could be assured by retaining invulnerable, de-alerted forces in place — specifically, missile-carrying submarines. To initiate the de-alerting process, Blair suggested unilateral transformation of ballistic-missile submarines to a low alert level. American nuclear submarines carrying close to 1500 warheads could be placed on a low, "modified" but secure, level of alert — removed from launch-ready hair-trigger status. A potential benefit of voluntary de-alerting might be to "immediately bring other nuclear-armed countries into the dialogue." In any event, proposals to eliminate spontaneous military reactions require involvement of the other nuclear-weapon states.

General George Lee Butler agreed that "we can achieve the immediate needs of arms control in large measure simply by taking the weapons off alert."

Also, by upgrading command and control systems and early-warning networks, greater resilience under attack can be achieved, allowing reduced reliance on prompt launch. Russia has been

> investing scarce resources to excavate deep underground command posts and upgrade an unusual second-strike command instrument formally called "Perimeter" and colloquially known as the "dead hand." If top Russian leaders do not get a clear picture of an apparent missile attack, or if for any reason they fail to give timely authorization to retaliate, the General Staff can activate this system to ensure quasi-automatic retaliation in the event of their decapitation.
>
> The latter ensures "quasi-automatic retaliation" that could be activated by the General Staff if the Russian nuclear-control system were "decapitated."

Reciprocal initiatives in peacetime alert status were advocated in *The Nuclear Turning Point*. A nationwide campaign to promote missile-launch de-alerting was initiated at the end of 1999 by "Back from the Brink," a coalition of arms-control groups.

Launch Impediments. Obstacles that physically (though reversibly) impede spontaneous missile launch have two advantages over de-targeting and de-alerting: Physical constraints are not easily restored, and they can be arranged so that the actions required to undo them will be visible. For example, removing umbilical straps or other wiring from a missile is effective in delaying launch, and its reversal would be conspicuous, at least to personnel assigned to monitor such activities.

For underground missiles, welding shut the silo doors introduces not only a longer delay, it also becomes suitable for remote monitoring; similarly, a large mound of dirt covering a silo would oblige a verifiable time delay in launch preparations. A comparable restraint for land-mobile missiles would involve disabling the transporter-equipment-launcher (TEL) or separating the missile from the TEL. Unused missile launch-tubes on submarines could be welded shut, and submarines' oceanic patrol areas could be restricted so that it would take days to sail to within range of targets.

Separating rockets from launchers and de-fueling liquid-fueled boosters are other missile flight-prevention practices that can probably be confirmed by on-site inspection.

Sandia weapons lab once compiled a list of factors that might affect agreement on unilateral or negotiated measures for missile-launch constraints. The considerations included: perceived threat to strategic forces, existing strategic balance or imbalance, purpose of strategic force, size and location of nations involved, resource limitations, and verification confidence.

De-Mating. Mentioned along with procedural de-alerting measures, the distinct physical action of de-mating involves detaching warheads or other key components, — such as computers, guidance packages, batteries, or igniters. It is the third major step recommended (assuming balanced and reciprocal measures).

Warhead "de-mating" can significantly delay restoration of a missile to operational condition; reinstating the system might take hours or days.

For major system components, simply increasing physical separation to storage locations well away from deployment areas or launch sites is a decisive step: return to service might take days or weeks. Even more durable is the actual disassembly of delivery systems or warheads: reassembly might take weeks or months.

An important virtue of physical separation is amenability to verification, either on-site or remote.

For long-range ballistic missiles, removing re-entry vehicles (that contain nuclear warheads) from the upper stage would constitute a readily verifiable form of warhead de-mating. The RVs could then be placed in secure storage, either nearby or at considerable distance: the greater the distance, the longer the delay in rearming and the better the reliability of verification. Such separation would be verifiable in any or all of three ways: by observing the de-mating process, by inspecting the storage facilities, and by periodically inspecting the missile upper stage. To provide continuity between inspections, overhead reconnaissance could supplement ground-based observations.

Former CIA Director Stansfield Turner proposed a kind of "strategic escrow": removing warheads from the missile and storing them up to 300 miles away, with assigned observers or remote monitoring to increase confidence in the separation.

Although de-mating the crucial components might be relatively straightforward, their return to service would admittedly require proven solutions to questions about operational safety, physical security, and quality control.

One important aspect of de-mating is a marked reduction in near-term risk of accidental or irrational nuclear danger without jeopardizing long-term national security.

Verifiability. For arms accords that apply physical constraints on missile launch, verifiability would be a major requirement. The thorough Sandia study mentioned earlier evaluated some possible verification techniques, taking into consideration strategic-force security (vulnerability and disclosure), practicality (cost, effort, equipment, facilities), effectiveness (success or circumvention), and launch-delay time.

Procedural verification measures — such as deployment restrictions and on-site inspection — are usually inadequate by themselves, so a combination of procedural and technical means would probably have to be invoked. Moreover, disclosing locations of mobile land- and sea-based missiles might reduce strategic-force security. Submarines, for instance, depend on location-uncertainty, so monitoring physical launch constraints on SLBMs would have to be done in a way that does not divulge specific boomer patrol areas.

Land-based missiles differ in that regard. One of the advantages of launch barriers for ICBM silos is that they could be made routinely observable by satellites.

Separation of missiles from their launchers could be continuously monitored by satellites. For land-mobile intercontinental missiles, the verification of TEL separation would have to be governed by agreed display schedules for observation by NTM and/or on-site inspection. For land-mobile missiles that are normally garrisoned in shielded garages (a Russian practice), various verification measures could be instituted, based on agreed protocols for routine stationing, maintenance, training, and exercises.

A previously mentioned way around the problem with submarine missiles is to weld the launch tubes shut; in an emergency, the welding beads can be removed. Missile tubes for SLBMs could be checked when the boat returned to port (intrinsic three-dimensional "fingerprint" castings can be made to verify weld integrity). An even more convincing launch constraint for home-ported boomers is to remove and store missiles without eliminating their capability to be reloaded.

Separation (de-mating) of RVs from missiles would be easier to verify than other types of physical constraints. (This is not the same as "RV de-MIRVing" where the number of warheads installed in an RV is decreased.) The fact that RVs had been detached and remain stored elsewhere could be checked periodically by on-site inspection, and probably by remote means if nose cones were left off the missiles. The extraction of critical components from missiles or warheads could conceivably be monitored remotely by tamper-resistant electronic seals that transmit to transponders on satellites.

Better yet, from the viewpoint of forced delay, would be the disassembling of warheads or delivery systems. Storing removed components at designated central facilities would simplify verification or monitoring, although with added vulnerability. Warheads disassembled and placed in storage could be verified by examining tamper-proof tags or intrinsic fingerprints on the casing.

In the past, difficulties in safeguarding design secrets while allowing inspection have been declared to be barriers to arms control agreements; however, with suitable political will, little if any sensitive information need be compromised. For a bilateral arms control agreement, there are probably no secrets that need protection between two militarily equal parties. For a multilateral treaty among nuclear-weapons states, different limitations or concessions are involved; for a still-broader international treaty, other safeguards need to be instituted. In all verification regimes, for whatever it's worth, the inspectors are sworn to secrecy and subject to sanctions.

Restrictions on Long-Range Missiles

Very short flight time is a pivotal and inherent characteristic of ICBMs and SLBMs, the type of weapons crucial for surprise nuclear attack or fast launch-on-warning response. Severely challenging the national-security calculus is the nominal half-hour intercontinental transit time — presenting the potential for a knockout nuclear holocaust with little realistic prospect of defensive interception. Nuclear-armed ICBMs and submarine-launched cruise missiles engender recurring nightmares in peacetime. Human launch-decision times cannot be much more than 10 minutes, and the time left to override the decision is just a little more prior to launch.

While considerable security is associated with ICBMs in silos as a deterrent, insecurity becomes personalized when you realize that opposing warheads are targeted at your own homeland. Ambassador Dean once succinctly expressed the continuing post-Cold-War dilemma:

> ... long-range ballistic missiles remain the crucial component of surprise nuclear attack and of the continuing danger of launch on warning....

Simply put, a mere accident or incident could precipitate mutual slaughter. Both parties would suffer nearly unimaginable and irreversible harm to people, their possessions, and their homelands.

Part of the unease over ballistic missiles comes from the danger of unauthorized or accidental launch with no recall capability. Once launched, it's too late to change your mind.

Fortunately, there are secure alternatives that could be implemented in a peacetime environment: The strategic network for nuclear deterrence consists of more than the temperamental ballistic and cruise-missile delivery systems. Submarines at sea, currently invulnerable, do not have to launch on warning. And (from home-based airfields or forward-based carriers) nuclear-armed aircraft, both strategic and tactical, can reliably and independently deliver weapons to targets. Compared to ballistic missiles, the relatively slow-moving aircraft have a much greater safety margin in terms of control and recall.

Proposals have been made for step-by-step missile limitation and reduction. First would be a negotiated global treaty to limit production and deployment of long-range missiles (exempting rockets designated for peaceful space activity); second would be increased transparency and data exchange on production and holdings; and third would be an agreement setting overall limits on holdings and production. Also, possible restrictions on cruise missiles could be considered. Such proposals for quantitative restrictions are in addition to suggestions addressed previously about missile-launch constraints, which apply to existing operational systems, and are directed at increasing the time required to react to a presumed provocation.

One of the most menacing U.S. missiles was the silo-based MX (LGM-118A Peacekeeper). Over 100 of them were produced, armed with 10 warheads apiece. Congress limited the deployment to 50 Peacekeepers. The total combined firepower for all 114 ICBM's was rated at around 342 megatons, or 342 million tons of TNT (approximately 2000 "Hiroshimas"). Because of their enormous weapons payload, they would be prime targets in a pre-emptive first-strike.[*] That made them the highest-priority candidates for conversion to single-warhead missiles at the earliest opportunity. The missiles were gradually retired, with 17 withdrawn during 2003, leaving 29 missiles on alert at the beginning of 2004. At the start of 2005 only 10 remained, scheduled to be retired by the end of the year. The last Peacekeeper was removed from deployment on September 19, 2005

While Trident submarines also carry MIRVs, the boomers are less vulnerable when at sea; even so, conversion to single-warhead missiles would diminish their threatening nature and reduce their value as targets when in port.

Verification of single-warhead loading would comparatively straightforward and reliable, especially for upper stages with narrow missiles.

Because reconnaissance satellites could be attacked by using specialized ballistic missiles, ICBMs are "a growing threat to the space-orbiting observation and communication satellites on which the international community is increasingly dependent."

Missile defense is driven by the (real and perceived) proliferation of long-range missiles and their capabilities. If missile proliferation could be controlled through collective international action, the problem of defending against them would be eased considerably. The Missile Technology and Control Regime (MTCR) has simply been an interim body hampered by its voluntary nature because of the self-interests of many members and nonmembers. In order to greatly ease not only the need for missile defense, but also the incentive for offsetting offensive capabilities, collective international action is needed, possibly leading to a treaty that provides verifiable strategic-missile production and deployment limits.

De-Creasing:
Transition to Low Levels of Warheads

At its zenith, the Cold War had motivated the production of more than 60,000 nuclear weapons; at one time, 5000 were being made in the United States every year. The arsenals settled at a total of about 50,000. By the close of the millennium, the combined number had leveled off to about 30,000.

Since then the decline has been slower, perhaps just a few percent per year, with no replacements being produced. Normal attrition allows the United States to dismantle about 1500/year, and Russia perhaps 2000/year. For 2003, the number of nuclear warheads remaining was estimated to be 27,000.

---

[*] A single nuclear warhead striking an MX silo would destroy ten warheads — a ten-for-one ratio. But if MXs had only one warhead each, their preemptive threat is reduced. Launch-on-warning need no longer be an urgent retaliatory policy.

It is said that Soviet-designed nuclear warheads have a shorter shelf life than those manufactured by the United States, which would explain why Russia keeps a significant reserve and routinely refurbishes its existing stockpile.

Most of the time, governments — echoed by journalists — refer only to the number of "strategic" weapons, leaving out the more plentiful mixture of "reserve," "tactical," and "stored" warheads. Although a large fraction of (strategic) warheads are being removed from operational status — which is highly desirable because of their immense lethality and precipitous potential — the total nuclear inventory is decreasing at a much smaller rate.

While a framework for significant reductions under START III had been discussed, deeper reductions to 1000-1500 deployed weapons have been proposed by Russia. More transparent verification methods could accompany such reductions, including the disposition of fissile materials (discussed later in this Section).

Wolfgang Panofsky advocated a "stepwise, reciprocal reduction regime" leading to a level of "a few hundred" nuclear weapons on the part of the United States and also Russia. Ratification of START II would have limited strategic nuclear weapons to about 3500 warheads. START III, which has been linked to ratification of START II, would have brought inventories to less than 2000. However, since the beginning of the millennium, superpower arms-control progress has been stalled.

The United Kingdom has unilaterally reduced nuclear-warhead loads of its three Trident submarines. Downloading from 60 to 48 warheads was scheduled in connection with each Trident's programmed docking. Also, the UK did not plan to order any more D5 missile bodies from the United States, "beyond the 58 missiles already purchased."

Criteria for stable reductions of strategic forces include: survivability, crisis stability, low launch readiness, arms-race stability, safety/security, transparency, and tolerable cost. In order for these criteria can be reconciled *Nuclear Turning Point* authors favored staged reductions ("arsenal de-creasing"); it corresponds to my fifth recommended step. Their vision for staged reductions affects all strategic delivery systems.

Arrangements proposed for START III and IV are merely suggestive of the possible choices that could be made. Land-based ICBMs, in silos or mobile, would be restricted to single warheads. Ballistic-missile submarines, inherently survivable, would have a reduced inventory of missiles and warheads. Bombers, capable of being recalled and not well suited to carrying out a surprise attack, are retained. SLCMs, extended in range and evasive in trajectory, play a strategic role, although they have been withdrawn from deployment since the 1991 mutual initiatives.

The biggest step (dubbed START V in *The Nuclear Turning Point*) would involve reduction to 200 warheads for each of the five nuclear powers. Various choices would have to be made to suit each nation's indigenous interests, warheads being apportioned to land, sea, or air delivery options.

Britain, France, and China would be expected to become involved when American and Russian levels go below 1000 strategic warheads.

Tactical nuclear weapons pose a separate problem; no arms-control treaty yet governs their inventory or deployment. To set the stage for shrinking these inventories, *The Nuclear Turning Point* advises data exchanges, withdrawal to storage, formalization of existing initiatives, and implementation of cooperative verification. Eventually the number of tactical weapons could be merged with the allowed number of strategic weapons. Cooperative verification measures include various forms of "fingerprinting" each individual warhead, so as to strengthen chain of custody through dismantlement.

In order to make transitions to smaller inventories, policy changes emanating from the White House need to be made. Top-level decisions on war planning are implemented via Presidential Decision Directives that set the standards for the SIOP. Moving from a "Major Attack Operation" as the prevailing strategy to minimal deterrence could easily reduce the strategic target list by a factor of 10, decreasing warhead inventory requirements accordingly.

Even deeper cuts — toward prohibition, abolition or renunciation of nuclear weapons — would be a logical, but distant end-state for progressive warhead reductions.

Economic Pressures. National economies are constantly under diverse pressures from the military and domestic sectors. In times of comparative global stability, domestic needs tend to gain ascendency. This is particularly the case for the nuclear-weapon states, who have sidetracked a half-century of resources to their arsenals.

Traumatic market-economy conversion is making it very difficult for Russia to maintain a large nuclear arsenal. Because weapon-system security and quality must be sustained, these are costly devices.

In 1998 the Pentagon had asked for cuts in U.S. expenditures for nuclear arms:

Driven by budget constraints as much as diminishing security threats, Pentagon officials are quietly recommending that the Clinton administration consider unilateral reductions in the nation's nuclear arsenal.

Because the U.S. has committed itself to drastic cuts in its nuclear arsenal, the Pentagon believes that the unilateral reductions would have no effect on America's ability to deter a nuclear adversary.

Because of Russia's delay and U.S. legislation blocking unilateral cuts, the Pentagon faces the prospect of paying hundreds of millions of dollars to maintain and rebuild weapons that the U.S. has agreed to scrap.

The Pentagon has spent $95 million more over the last two years than it would have if START II had taken effect. It would cost $100 million more next year and $1 billion the year after that.

Both the UK and France have embarked on unilateral reductions in nuclear arsenals. Of course, they would like to see the larger powers make comparable parallel reductions.

De-MIRVing. As part of the transition to low levels of missiles and warheads, my fourth recommended step is "de-MIRVing" — reducing to one RV per missile. (I prioritize it ahead of "de-creasing" arsenals because it has its own beneficial dynamic for strategic stability.)

Reducing the number of RVs on each missile could encourage reversion to the type of more stable deterrence that existed before the SALT treaties eliminated constraints. No self-control at all was exercised in ballistic-missile MIRVing after that, but some forbearance in developing defensive systems took place, as reflected in SALT-I ABM treaty.

Failure to tame the multi-headed MIRV monster still haunts the process of arms reduction. Large, special-purpose missiles were built just to carry multiple warheads. This created an inherent and enduring lethargy against reversing the "bigger and more" process.

Major decrements, rather than single-warhead reductions, are now needed to accelerate peacetime progress toward nuclear stability. Russia's Topol-M single-warhead missile is a shift in the right direction, one that other nuclear-weapons states would do well to emulate.

After de-MIRVing, other possible stabilization steps could follow, such as de-mating and disassembling RVs. The separated components could be retained near the missile or in distant storage, and the warheads could be kept intact at that phase of the transition. These steps can be witnessed without revealing (to another nuclear-weapons state) sensitive information. Moreover, the warhead — upon removal from its RV — could immediately be tagged or fingerprinted for future tracking. Bilateral, multilateral, or international monitoring can be organized for stored warheads or RVs.

Without undercutting effective deterrence, de-MIRVing by itself reduces the other side's incentive to launch a pre-emptive attack. It also significantly diminishes the consequences of retaliatory launches. All of these measures, especially those that go beyond de-MIRVing, implant delays that frustrate precipitous, unauthorized, or accidental launches. For long-term confidence in an agreed, mutually stabilizing stand-down of nuclear forces, de-MIRVing must be coupled with on-site verification and/or remote monitoring.

De-Fending:
Cooperative Missile Defense

Discussions of defense against long-range missiles have commanded attention for more than four decades, but the topic is still highly controversial. As described in previous Chapters, prolonged disputes have broken out about each of the proposed ABM, SDI, NMD, and BMD systems — none of which has been able to convincingly overcome technological and other limitations.

Nevertheless, persistent uncertainties exist about potential threats from long-range missiles launchable from distant nations. Some scenarios posit various political, strategic, tactical, and psychological challenges. As a result, active ballistic-missile defense has gathered reasoned and sometimes vociferous support, even after demise of the Cold War.

Some informed individuals interested in national security and arms control have not out-of-hand dismissed the potential benefits of ballistic-missile defense. For example, General Charles Horner, former commander of the U.S. Air Force Space Command, has explicitly acknowledged that ballistic-missile defenses could facilitate the reduction and eventual abolition of nuclear weapons.

However, in the absence of accompanying offensive-arms reductions, the deployment of missile defenses has encountered resistance. When the Clinton administration announced funding for unilateral deployment of a national missile defense (NMD) system, *Arms Control Today* wondered whether this announcement was

> initiating the next great arms race of the twenty-first century? With the proliferation of ballistic missiles, are we going to see the outright militarization of space ... the return big time of multiple independent re-entry vehicles?

Their answer to that rhetoric question about NMD was

> the more immediate impact will be the chilling effect it's going to have, or worse, on the ongoing process of moving away from nuclear weapons and continuing drastic reductions in nuclear stockpiles.

Another response (expressed by Susan Eisenhower) was that "this could set off a new arms race when there is really no immediate threat."

The alternative of a collaborative, rather than unilateral defense buildup is not new, having been suggested from time to time, but never seriously by policymakers. Explored here in more detail, is the feasibility of a BMD regime that would be an asset rather than a liability: It would present the salient features of a ***coordinated*** missile defense technically practicable and politically acceptable. The concept centers around missile defense undertaken by nations in partnership. This is the sixth, and perhaps most contentious step recommended in this book toward mutual and stable nuclear equilibrium.

The following paragraphs describe strategic and technical opportunities — offense/defense tradeoffs, multi-phase defense, and cooperative defense — as well as benefits in dealing with large- and small-scale threats. These constitute ballistic-missile "de-fending" to help bring about a normative post-Cold-War nuclear-arsenal phaseout.

One way to start this initiative would be to revive, supplement, and expand the ABM treaty, allowing periodic revisions and participation of additional parties, particularly other nuclear-weapons states. However, nations interested in a fully cooperative approach would now be better advised to start almost from scratch, drawing up a new treaty, using still-relevant parts of the ABM accord.

Consideration of a cooperative and feasible missile defense is based on several key coordinated processes: gradually exchanging offensive systems in favor of ballistic-missile defense, broadening the scope of missile defense to enable reaction during all feasible flight phases, and cooperating multinationally on agreed defensive measures.

Besides offering more realistic security against the growing threat posed by ballistic missiles under the control of many nations, one can anticipate that a cooperative defense structure would allow — will promote — deep reductions in nuclear arsenals.

Offense/Defense Tradeoffs. Development and deployment of a ballistic-missile defense could be tightly tied explicitly to agreed reductions in strategic missiles. If significant cutbacks take place in offensive-missile holdings, ballistic defense becomes more promising, especially if all nuclear-weapons states eventually participate in the process.

Under a (bilateral or multilateral) negotiated missile-defense treaty, favorable defensive concepts could be developed, tested, and deployed at a rate commensurate with reductions in offensive systems.

Simultaneously, or in tandem with missile-defense negotiations, discussions would proceed about reducing missile-delivery systems, with the objective of lowering the number of offensive missiles that need to be countered. The challenges of mid-course and terminal interception would be greatly simplified by agreed and verifiable de-MIRVing of missiles. Countermeasures against defense, such as decoys, could be banned. Cruise missiles are an illustration of alternative offensive delivery systems that could be subject to payload restrictions and to verification that they are not nuclear armed. Indeed, intrusive inspection would be central to a cooperative BMD regime.

A transition from policies of "offense dominance" to "defense dominance" could be eased by signing agreements that balance weapon systems. Reductions in strategic forces could be timed to coincide with improvements in offensive risk-reduction and active-defense buildup. This would lead to a strategic situation that could be termed "mutual assured security" (in lieu of mutual assured destruction).

As rhetorically posed by Alvin Weinberg, one of the pioneers of nuclear science, we need a "defense-protected" build-down, in which the key issue is a compensation ratio: What number of offensive missiles would be eliminated as defense effectiveness improves?

Multi-Phase Defense. To be comprehensive, a discussion about structuring a defense program would tackle every recognized missile-threat aspect. Some relevant topics — such as missile policy, strategy, posturing, proliferation, and production — have been touched upon earlier in this book.

Any practical approach to missile destruction must reckon with three flight phases — *boost, mid-course*, and *re-entry* (or *terminal*) — during the nominal 15-30 minute intercontinental trajectory. After liftoff, a missile is powered (boosted) out of the Earth's atmosphere, into mid-course ballistic flight in space. That trajectory continues until the terminal phase, when warhead-laden RVs separate from the bus and re-enter the atmosphere en-route to their respective targets. For each of these three flight phases, defense planners assign different "tiers" or "layers" of defensive apparatus, including means for detecting launches, tracking missiles, and sensing stage separation.

Ballistic missiles can be armed with multiple RVs, some of which might have defensive countermeasures to complicate interception during mid-course and terminal phases. Because it is difficult to intercept mid-course RVs immediately after they separate from their missile bus, any comprehensive and practical "layered" ballistic-missile defense emphasizes an earlier *boost-phase* defense — as problematic as it might be.

Interception at the boost phase, which occurs soon after missile launch, demands three successful preparatory actions: (1) instantaneous sensing of rocket firing, (2) rapid (first few minutes) assessment and decision to intercept, and (3) interceptors that are available, proximate, and very fast.

If it could be arranged as part of a cooperative agreement, one way to finesse these technical challenges would be able to carry out remote-controlled destruction of rockets at or during liftoff. A negotiated arrangement of that type has the distinct advantage of circumventing technical problems in discriminating and intercepting multiple targets in later flight stages.

If remotely controlled missile destruction is not implemented or does not succeed, the next defense layer would be boost-phase intercept. For boost-phase kill (by interceptor missiles stationed on land, at sea, or in space), autonomic reaction must occur within the two- or three-minute window between ignition and stage separation. Also required for success would be proximity of an exceedingly fast and responsive means of tracking and intercepting, all daunting problems. Clearly the reaction and intercept process would have to be automated, which might only be feasible through a cooperative process (that included built-in remote-destruct mechanisms).

Among the challenges of boost-phase intercept are brief liftoff time, high-acceleration needed for interceptors, unpredictable ICBM flight acceleration, and payload-disabling complications.

Another option, if allowed by cooperating partners, would be high-power lasers that are airborne or satellite-based and routinely kept on station within range of missile-launch sites.

Many arguments for and against boost-phase missile defense, as well as the technical prospects and deficiencies, are detailed in *The Phantom Defense*. Some, boost-phase defense deficiencies, perhaps enough of them, could be overcome with a negotiated cooperative defense.

Mid-course defense has for intercept the most time available: about two-thirds of the nominal 10- to 20-minute flight time. RVs from MIRVed missiles begin their separation from the bus during this phase. The U.S. NMD program concentrated its testing on mid-course interception using ground-launched kinetic-kill vehicles, trying especially to discriminate warhead RVs from dummy objects.

Space-based anti-missile platforms would be particularly valuable for mid-course ambush. Nevertheless, cooperative defense would circumvent the need (and associated problems) of detonating nuclear warheads in space in order to have a high probability of missile kill.

Terminal defense, to be activated when RVs re-enter the atmosphere near the target, is a last-ditch measure. Because of disproportionate masses, warheads separate from decoys by "atmospheric sorting." This interval permits a defense concentrated on incoming warheads. Although terminal defense offers a better technical opportunity to discriminate warheads from decoys, it cannot defend large regions — only selected, high-value targets.

The long-standing Galosh ABM system around Moscow, inherited by Russia from the Soviet era, was a desperation-style terminal defense relying on atmospheric nuclear explosions to destroy or disable incoming RVs.

As explained below, the three conceptual phases of missile defense could be assimilated into a coordinated intercept system, thus compounding the potential value of a negotiated cooperative defense.

Cooperative Defense. Cooperative development and coordination might remedy many inadequacies of unilateral ballistic-missile defense. During the Cold War, both adversarial superpowers naturally went their own way.

Now, however, agreed joint undertakings could effectuate measures such as pre-launch-control enhancement, post-launch self-operative destruct, and automated boost-phase interception. For mid-course and re-entry phase defense, these augmentations would alleviate weaknesses or supplement capabilities of independently developed national technologies. A comprehensive cooperative approach to missile defense — though no political shoo-in — offers greatly increased prospects of successful interception.

Cooperative defense substantively diminishes many weaknesses of unilateral ballistic-missile defense.

The following four boxes outline the negotiated missile-defense concept, describing lift-off control, automated destruction of unauthorized missiles, national BMD, and cooperative BMD. Indigenous national missile defense would be similar to the framework that's been under development for many years in the United States, except it would benefit significantly from related features provided by a cooperative missile defense. Some limitations are discussed in subsequent Subsections; it is not a cure-all.

[Negotiated Missile Defense – Lift-Off Control]. Integral to cooperative defense is the ability to formulate pre-launch preventive actions which increase assurance that a particular missile to be lofted is not armed with a warhead. We're talking about peacetime conditions, when pre-launch payload inspection by intrusive or non-intrusive means could be carried out as part of an agreed verification process. The capabilities of non-intrusively verifying the absence of a nuclear warhead are well in hand.

International management of civilian and military missile firings would be conceivable with coordinated central control. Analogous to regulation of aircraft flights and landings, pre-approved itineraries and pre-launch clearances could be required. A launched missile not cleared in advance could be subject to automated interception and destruction. Existing infrared-sensing satellites can be used to detect heat from missile exhausts within seconds of a launch. Under a cooperative agreement, liftoff-sensors could be situated at the launch site.

Missile-launch clearance would be based on satisfaction of several criteria, such as the filing of a flight plan, the payload capacity of the booster, and approval by pre-inspection. Because rocket firings are comparatively infrequent, there is plenty of lead time to carry out clearance procedures without delaying lift-off preparations and schedules.

[Negotiated Missile Defense – Automatic Destruction]. Unless a ballistic-missile flight is preauthorized, it would be automatically destroyed by any of the several mechanisms put in place through a negotiated-defense treaty.

By requiring prior missile inspection and launch clearance, the time-critical problem of deciding whether to automatically abort or deliberately destroy a rocket is vastly simplified and expedited. Moreover, unified control-center communication could quash the possibility of unwarranted retaliation initiated as a result of signals misinterpreted to represent a missile attack (as it mistakenly, but briefly appeared during the Cold War).

The Joint Data Exchange Center (JDEC), established as part of a June 2000 U.S./RF agreement on early warning systems and notification of missile launches, was a stride in the direction of cooperative missile defense. The sharing of early-warning information, at the jointly staffed JDEC, was intended to provide a "near real-time exchange of the detected information about the launch of ballistic missiles and space-launch vehicles." Information to be sent to JDEC could come from each party's intelligence satellites and early-warning radars.

Pre-launch notification and authorization of launches is essential; later phases of a missile-defense regime will incorporate information and procedures regarding third-party launches. Expanding the missile-warning system to give worldwide coverage would be a major advance toward viability of a negotiated BMD. Nations that have not subscribed to the negotiated missile-defense regime risk having their ballistic missiles destroyed in flight.

[Negotiated Missile Defense – National BMD]. National missile defense is a central component of any cooperative system — not standalone, but coordinated with other treaty partners — yet self-sufficient as a last resort. Without this coordination, NMD is a weak and porous defense; with worldwide cooperation, NMD gains symbiotic strength.

The actual means of ballistic-missile intercept, during any flight phase, are to be derived from national missile-defense systems. Detection from land-based radars and space-based sensors could be a combination of national and international missile launch and flight sensors. (National intelligence sources would remain primarily beholden to domestic security.)

Space-based sensors, computers, and communications — but not weapons — are likely to be necessary in order to have an effective capability for re-entry-phase interception.

Mid-course defense would be augmented by agreed cooperative measures. Exchanging real-time data, especially from satellites and over-the-horizon radars, is essential for synchronization of defensive phases.

Meshing all three BMD phases — boost, mid-course, and terminal — greatly increases likelihood of intercept success.

Before the United States withdrew from the ABM Treaty, a plan to jointly develop, test and command any new ballistic-missile defense was considered by Russian military analyst Alexei Arbatov to be "essential," along with accelerating and upgrading current cooperation on theater missile defense, joint early warning systems, and command centers.

[Negotiated Missile Defense – Cooperative BMD]. Integral to a cooperative system is each nation's agreed contribution to missile defense. Combined and coordinated, these national systems constitute a more effective universal means of mutual defense.

Separating civilian from military missile-launch sites would reduce the need for land-based boost-phase interceptors having to be stationed near civilian sites. In addition, non-military missiles that undergo pre-launch payload verification would not have to be subject to intercept.

National missile launch and tracking systems would be adapted for cooperative benefit. International radar arrays and satellite sensors could supplement national missile-detection systems.

Treaty provisions might permit phased-array radars in an ABM mode and practice shots against boosters and re-entry vehicles launched by a cooperative party. Such an agreement would probably forbid anti-satellite systems and the stationing of defensive battle stations in Earth orbits. The use of nuclear explosives for defensive purposes would probably not be acceptable; however, development of ground-based laser, particle-beam, and kinetic-energy weapons might be allowed.

The Brookings Institution *Nuclear Turning Point* recognized the value of cooperative defense, which could take many forms: information exchanges, technology sharing, joint development of defenses, and even collaborative defense deployments. Moreover, with the inclusion of China, they suggested:

Early warning systems would seem to be particularly promising for such cooperation.... A U.S. commitment to worldwide cooperation on missile defenses, perhaps beginning with a global early warning system and leading to standardized missile defenses available to all participating nations, could reduce concerns.... The deployment of an internationally controlled ballistic missile defense system ... might similarly be less threatening if the difficult problem underlying how to deploy and control it could be resolved....

But *The Nuclear Turning Point* also predicted that extensive cooperation on such defenses might be "improbable" and added their concern that it alone it could not solve "the underlying problem: defenses that threaten to nullify a country's deterrent will provoke offensive reactions."

That's partly true, but a well-considered negotiated defense should be able to circumvent such feared limitations in assured retaliatory capability. For instance, a negotiated missile-defense plan could enable use of an international ballistic-missile PAL (Permissive Action Link) — a device that interlocks ballistic missiles within a multinational electronic-security system. Only under certain defined (but unilateral conditions) would a government be able to release one or more of its land-based missiles for launch.

Aircraft bombers and submarine-launched missiles are likely to linger on, so land-based ballistic missiles would be simply part of overall strategic offense. Sea-based missiles, which cannot be easily neutralized by any ballistic-missile defense, are considered to be less of a risk for unauthorized or accidental launch. However, ICBMS of nations that are outside the cooperative-defense arena are far more likely to be land-based.

Cooperative pathways could erase the unnecessary dividing line between unilateral homeland missile defense and a united international missile screen. The unilateral approach is inherently deficient because there are too many possible threats. The international system, being multi-layered, would have a greater prospect of success for participating parties. Previous Russian and European reservations about ballistic defense would be alleviated by an inclusive jointly developed international system.

The May 2002 Moscow summit ended with agreement on the steps toward missile-defense cooperation that brought JDEC into operation, set up a bipartite observation-satellite program, arranged for joint modeling and simulation, and initiated mutual technology cooperation:

[The] United States and Russia have agreed to implement a number of steps aimed at strengthening confidence and increasing transparency in the area of missile defense, including the exchange of information on missile defense programs and tests in this area, reciprocal visits to observe missile defense tests, and observation aimed at familiarization with missile defense systems. They also intend to take the steps necessary to bring a joint center for the exchange of data from early warning systems into operation.

The United States and Russia have also agreed to study possible areas for missile defense cooperation, including the expansion of joint exercises related to missile defense, and the exploration of potential programs for the joint research and development of missile defense technologies, bearing in mind the importance of the mutual protection of classified information and the safeguarding of intellectual property rights.

> The United States and Russia will, within the framework of the NATO-Russia Council, explore opportunities for intensified practical cooperation on missile defense for Europe.

A Russian-American early-warning data-exchange center could begin pointing the way to joint launch regulation and managed interception of errant missiles.

Success would depend strongly on cooperation. With coordination, teamwork, and collaboration between the two major nuclear powers, the prospects of their respective missile defenses would significantly improve. Participation by other nations will benefit all, especially if defensive systems become internationalized. Those nations or entities that keep themselves outside of the negotiated missile-defense structure would present reduced low-probability, small-scale threats, which are discussed next.

What must be kept in mind is that missiles will be forevermore frequently launched, and fear about their capability will be ever-present, with or without national or mutual defense. Much, though, can be done meanwhile to manage nonautonomous risks.

Small-Scale Threats. So far, it has not been necessary to be specific about the number of missiles that have to be countered. Two threat scenarios, limited to a single or a few missiles but still of high consequence, are (1) accidental or unauthorized launches from existing missile-capable states and (2) deliberate (provoked or retaliatory) launches from so-called "rogue" states, perhaps in response to an invasion or reaction to "surgical" strikes against their nuclear facilities. The declared primary focus of the Bush-administration BMD scheme was defense against "rogue" states.

In severity and quantity, these small-scale threats differ from a large-scale, all-out attack that might stem from a deliberate (or irrational) decision by a well-armed adversary. Needless to say, any ballistic defense is likely to be easily overwhelmed by a large-scale attack. However, the technology for blocking a small-scale attack is far more promising, especially if cooperative defense were widely adopted.

To illustrate some possibilities involving small-scale threats, let's consider how cooperative missile defense would cope with them. To deal with possible unauthorized or accidental launches (from designated sites) of treaty parties, override devices could be installed in rockets by mutual agreement. The overrides would defuse warheads or destroy boosters automatically. (Of course, all nations would preserve the right to bypass devices in a time of declared emergency.)

In protect against non-signatories (non-cooperative parties), participating nations would retain their national BMD capacity to intercept uncleared (unauthorized, accidental, or deliberate) launches wherever they might originate. Moreover, certain non-signatory nations would become the focal point of boost-phase defense installations, thanks in particular to the collaboration of neighboring signatory nations. Clearly, the greater the number of treaty participants, the more effective the common defense. Although a National Academy of Sciences study ascribed little technical credibility to boost-phase interception, they did not mention if detailed attention was given to what could be achieved with a negotiated missile-defense regime.

Any nation, including a missile-defense partner, that does not get its launch schedule cleared in advance, even for scientific missions, would risk having its rockets reflexively destroyed upon launch. Considerable incentive would exist for obtaining advance clearance, at least on an *ad-hoc* basis.

Boost-phase defenses would become more viable against "rogue" nations that remain outside negotiated offense-defense agreements. Not only would existing detection and interception technology be more adaptable to boost-phase defense, but deployment sites could be situated where they would be more effective, e.g., stationed on ships off-shore or land-based on adjacent foreign territories (as entitled by a multinational accord). The Obama administration has indicated its serious consideration of Mid-East-area intermediate-range missile defenses that include partner nations.

In addition to the boost-phase measures mentioned above, treaty parties would develop and retain mid-course and terminal defense, both of which would be primarily national choices but augmented by inter-nation cooperation in missile detection and destruction.

A useful joint-defense treaty need not begin with multilateral participation; to get things going it might simply be bilateral collaboration between the United States and Russia. In 1990 Gorbachev proposed the "development of joint ABM early-warning systems to prevent unauthorized or terrorist-operated launches of ballistic missiles." Although this proposal was intended in part to deflect European criticism of the USSR, it provided an opening for Soviet policy entrepreneurs who in any case wanted to promote the development of strategic-defense systems.

Richard Garwin has recommended that the United States negotiate with Russia, Ukraine, Belarus, and Kazakhstan to permit deployment of boost-phase interceptors that could be placed close to North Korea and Iraq. Garwin also suggested that cooperation with Russia could reduce the risk of possible accidental launches.

To intercept an accidental or unauthorized launch from Russia, the United States could have a missile-defense installation in the northern Midwest states or in Canada. To intercept a missile from North Korea, a joint U.S.-RF agreement could allow a site to be deployed on Russian territory south of Vladivostok. Against Iraq, a single interceptor site in southeast Turkey would have been enough to protect the route to the entire continental United States.[*] For the Mid-East, combined sea- and land-based defenses would be appropriate.

Through the coordinated implementation of a negotiated, multilateral treaty on missile offense and defense, small-scale ballistic-missile threats could be significantly diminished.

A negotiated defense initiative could readily expand from bilateral to multilateral because most, if not all nations have a common interest in averting nuclear wars,

---

[*] It was not a credible idea that post-millennium Iraq would have had either the ability or motivation to launch an unprovoked attack against the continental United States.

denying second- or third-party threat capability, and intercepting unauthorized or accidental launches. The NDI is a practical initiative that strong-defense advocates should favor.

De-Weaponizing Outer Space. It follows that nations interested in cooperative defense would want to protect their space orbiting assets from co-option. In Section B particularly the counterproductive role of ASAT weapons was discussed. These, and other space-based weapons, would be banned from Earth orbit.

By "de-weaponizing" space (seventh step in my comprehensive list), land- and air-based ASAT capabilities would be subject to negotiation depending on their limitations. Development of launchable ASAT would be restricted.

Although weapons in space are already banned by the Outer Space Treaty, other orbiting devices that might negate or complicate missile defense might also have to be restricted.

Outer space is not necessarily the only common jurisdiction or realm in which strategic defense is improved if certain types of weapons were to be banned.

Strategic-Defense Enhancement. Criticisms have been enumerated for the unilateral NMD plan under development by the Clinton and Bush administrations; many opponents fear that the plan would not be conducive to reductions in offensive missiles or that a new military competition would be promoted.

Such lingering fears are a major justification for a negotiated defense initiative. By having agreed reductions in strategic offense, there would be fewer missiles and RVs for the defense to cope with. Thus, cooperative measures make the technology for defense more feasible, especially for prompt, correct decisions and reactions within the short time window for recognizing valid missile threats. Defensive systems, facing fewer projectiles, would thus make a viable contribution to mutual security, enabling iterative reductions in excess strategic offense.

By reducing the number of weapons that have to be countered, the cooperative approach does not have to compensate — as it must do now — for the porosity of a unilateral defense deployed to face huge offensive arsenals.

Arms stabilization — however achieved — averts dangers of future conflict escalation. This is particularly important now because the superpowers still have overwhelming nuclear arsenals and the means to deliver them.

Conversely, strategic-arms reductions inevitably improve the prospects and perception of successful cooperative defense.

Perhaps the inventory of retaliatory missiles would have to drop to the hundreds or tens before the benefits would be fungible, but negotiated ballistic-missile defense provides an interim incentive for quantitative reductions. In the meantime, decades have passed and many more are ahead before the superpowers will divest themselves of a large fraction of their nuclear arsenal without added incentives.

Defensive capability would also be enhanced if agreement were reached on banning countermeasures, such as decoys. Such a prohibition could be verified through RV-inspection protocols written into the treaty.

Investing in cooperative missile defense results in technology-sharing, making it less expensive for each participant. This frees up defense resources that could be allocated to defeating alternative means of nuclear-weapons delivery. Neither the negotiated nor unilateral missile defense schemes would be able to directly solve vulnerabilities related to non-ballistic means of delivering weapons of mass destruction or indiscriminate casualty; however, by establishing international collaboration against one particularly abhorrent peril (nuclear-armed ballistic missiles), a precedent could be set for joint endeavors against other threats and means of delivery.

In-flight "adversarial" testing could be one useful outcome in an early stage of cooperative-defense development. Improvements in BMD feasibility and strategic stability would create positive feedback to preserve and strengthen the arms-control process.

Partnership among nations would finesse many objections to anti-ballistic-missile defenses. Rogue states would be factored into a broader plan dealing with all nations that have long-range missiles.

Cooperation in missile defense suggests other benefits: Better use of existing technology when compared to the lack of existing workable technology for solo defense and the undertaking of cooperative ventures. Abrogation of the ABM Treaty and premature deployment become moot issues because negotiated defense would encompass new conditions and schedules. Negative public and foreign reactions to missile defense might be turned around, and research and development conducted to date could be the starting point for more a comprehensive and effective ballistic-missile shield.

Besides making missile defense technically more feasible and politically more acceptable, a negotiated initiative should make it more cost-effective at the margin by spreading the research and development among partners and by being more likely to succeed.

Japan has sought to purchase from the United States a land- and sea-based defensive system to use against intermediate-range missiles. Such activity might bother China, which does not want missile defense extended to Taiwan. This is another example of the need for broader collaboration in defensive measures.

Aside from the direct improvement in missile defense, there are collateral benefits that could accrue; one of these is that the defense program could lead to joint commercialization of relevant technologies, including some originated during the Cold War.

Moreover, by organizing technical collaboration on ballistic-missile risks, R&D could be extended against a common threat to humanity: asteroids that penetrate the Earth's protective atmosphere. Some of the technologies useful for multilateral missile defense are applicable to detection, warning, and reaction to an external asteroid threat. Measures to deflect or shatter an asteroid would require science and technology adapted from national defense and space programs.

In short, negotiated missile defense, global and harmonized, could improve national security. Combining reductions in offensive forces with cooperative development of missile defenses could lead to replacing MAD with restraint-dominated mutually-assured security.

De-Militarizing:
Warhead Destruction

The overall objective of warhead dismantlement under a deep-cuts regime — without nations actually giving up basic weaponization expertise — is to delay the time required to reconstitute nuclear armaments *in extremis*. A realistic expectation is to prevent rapid breakout and to delay reversion to massive nuclear confrontation. These are achievable goals.

It's not going to be easy to persuade long-standing and embedded military, industrial, governmental, and scientific interests within nuclear-armed nations that deep-cuts are feasible to achieve and do not undermine national security. That's why substantive, realizable objectives are needed: to overcome institutional resistance.

If deep-cuts were mandated, nuclear-armed states would relinquish weapons, but not the knowledge or capacity to rebuild. Not much more than that can be realistically expected. That's one reason why "abolition" isn't a credible goal at the present time.

To impede nuclear backtracking, it is necessary to introduce a tangible and durable recovery lag — measured in years rather than days. Diplomatic processes will need time to resolve tensions without nuclear confrontation.

Mutual confidence can be built up gradually as de-militarizing goes forward (the eighth of my nine recommended steps). It would take decades, rather than years to dismantle and destroy thousands of warheads and their fissile materials. A practical objective would be to build in enough institutional and technological barriers that nuclear rearmament would take at least a year or two.

Although destroying only the warhead-delivery vehicles might seem sufficient, the nuclear payloads themselves warrant destruction because they pose both a domestic burden and an international threat. Few nuclear weapons would have to be retained for deterrence; none are needed for homeland defense, because using them within one's own national borders would be self-destructive.

Much infrastructure for warhead dismantlement is already in place and being routinely exercised while existing nuclear weapons are retired. The post-Cold-War U.S. capability for warhead dismantlement was about 1500 to 2000 weapons per year, substantially smaller than the peak buildup rate of about 5000 per year. Formalized deep cuts would increase the pace of dismantlement, requiring additional facilities, funding, and qualified personnel.

Facilities that might become overloaded — on the critical path in a phased-reduction regime — would primarily be those involved in warhead dismantlement and materials disposition. Storage limitations might temporarily delay the process until enough weapons-assembly personnel and facilities were qualified to disassemble parts safely. The infrastructure for burnup (or burial) of separated fissionable material would need to be expanded and maybe subsidized.

Even if abolition advocates got their way, the ingredients to make nuclear weapons will unavoidably be with us for a great many years. That's why there is an ever-looming threat of resurrection. A pragmatic objective would be to introduce a comprehensive series of impediments that forestall quick reversion to warlike nuclear arsenals.

General Butler has warned that "we're going to have to live with large numbers of [nuclear weapons] for some years to come ... owing to the lack of facilities [for dismantling warheads]." In any event, "for the United States, the Cold War cannot truly be over until we face up to the nuclear question."

Disposition of Nuclear Materials. The act of dismantling nuclear warheads not only removes them from immediate availability for warfighting, it introduces a time lag before they can be reconstituted — a process that requires very special facilities and highly trained personnel. Section E contains a detailed description of nuclear-materials disposition options; just a summary is included in this Subsection.

Crushing or otherwise wrecking warhead components imposes immediate but reversible delays in recovery, because various specialized machining and fabrication processes are needed. A more enduring and practical step impeding reconstruction of nuclear arsenals is to degrade the fissile constituents by "de-militarization." By doing so, replacement-material production would become a substantial task measurable not in months but in years. Demilitarization takes place essentially by irreversibly denaturing fissile properties of the nuclear constituents.

The crushed nuclear pits by themselves would reveal few if any secrets, but the materials must still be safeguarded to prevent diversion. It would be appropriate to require the responsibility for enumerating and accounting nuclear materials at storage facilities to be shared by the national authority and an international inspecting agency.

[Global Stockpiles of Uranium and Plutonium]. The best public reckoning in 2005 of worldwide HEU stocks was about 1750 tons, 10 percent being in civil use. The figure for military plutonium was 155 tons. Estimated by NGOs such as the NRDC, these amounts are essentially unchanged over a decade, and are not like to meaningfully decrease until systematic demilitarization is implemented.

For civil plutonium the figure is 1700 tons, which will continue to increase as reactors produce more power for world needs. Only if plutonium recycling is implemented on a worldwide basis will this quantity diminish.

Civil plutonium and uranium are subjected to domestic and international safeguards. These reactor-grade materials are, for all practical purposes, immaterial to weapons and irrelevant to cuts in nuclear arsenals.

Individual nuclear warheads require merely kilograms of either or both HEU and military plutonium; so inaccuracy in nominal estimates of worldwide inventories does not really affect technical-policy issues about nuclear weapons.

Because of political momentum and technical inertia, the military inventories aren't expected to change much in coming years.

Irreversible disposition of enriched uranium can be accomplished rather simply, as described in Section E, by blending natural or depleted reserves. Plutonium, however, requires transmutation of its fissile isotopes, a process that would generate a profit rather than expense as the material is consumed as fuel in reactors.

[National Security Bonus]. The extremely rare isotope helium-3 is especially effective in neutron detection of possible plutonium smuggling. It's primary source is from decay of tritium used in nuclear weapons. If deep nuclear reductions were to take place, much more helium-3 could be salvaged upon dismantlement of warhead pits, thus helping to alleviate a serious shortfall in its supply.

Reduced production requirements for tritium and its increased availability for non-military applications would also be a bonus.

Unless these progressive atrophying actions are carried out, the notion of deep cuts in arsenals would be wishful thinking. Mechanical dismantlement of warheads is insufficient. Only by destroying warhead components and demilitarizing the nuclear ingredients does it become possible for both nuclear- and non-nuclear-weapons states to become comfortable with the disarmament process.

Defense Conversion

Converting nuclear-weapons complexes to nonmilitary enterprises would be consistent with a multifaceted strategy that gradually ushers in durable barricades against rejuvenation of nuclear arsenals. In the course of time, the facilities and personnel would lose first-hand capability to engage in rapid breakout from a disarmament treaty. They would still have the generic skills and institutional documentation to sustain minimal arsenals or reconstitute emergency weapons.

Resistance by military-industrial-laboratory interests would be expected to defense conversion (infrastructure "de-minishing," the last of nine recommended steps in nuclear de-emphasis), but the gradual retirement of Cold-Warriors and transformation of weapons labs to productive peaceful activities would soften the challenge.

Budgetary trauma associated with converting Russia to a market economy has made it necessary for Western nations to provide financial support for defense conversion — in addition to assisting storage, dismantlement, and disposition of former Soviet weapons and their ingredients.

Early on, the G.W. Bush administration announced post-election plans to re-evaluate the U.S. programs with Russia that were intended to stop the spread of nuclear, chemical and biological weapons. The *Chicago Tribune* welcomed the review, suggesting that the administration should reassess programs

> such as those dealing with nuclear materials and safety; implementing plans to destroy 50 tons of plutonium that could be made into thousands of weapons; adding to efforts to make Russian fissionable material secure from theft or loss; putting 8000 Russian scientists to work in the civilian research sector; tightening export controls in Russia, and helping Russia convert from chemical and biological weapons.

At home, there will be competition for those dollars, including a backlog of maintenance deferred at U.S. nuclear weapons facilities of the Department of Energy. Those programs also are important, but Russia's nuclear weapons are the greater threat.

It is in the U.S. national security interest to help dismantle Russian nuclear weapons. It is in the national interest to make sure that the dollars committed to achieve that goal are well spent on effective efforts.

While the Bush administration did reassess the nuclear-support programs, they did not immediately or fully heed the *Tribune's* warnings or recommendations.

Other aspects of infrastructure conversion could include limitations on nuclear testing based on adherence to the CTBT and NPT. For example, while nuclear "subcritical experiments" might be permitted for confirming the reliability of warheads retained during a drawdown of arsenals, opening up the process to outside verification would diminish the rationale for other nations to develop their own nuclear weapons. Without being committed to test restrictions, the G.W. Bush administration policies retained a nuclear-attack strategy, coupled with planning that had ambiguous implications. Together, these postures were more likely to stimulate than to tame nuclear proliferation.

Synthesis of Recommendations

Overall conclusions for this entire three-volume book are summarized in the Denouement that follows this Chapter. The recap is necessarily incomplete — a listing of all recommended ways to mitigate nuclear risk would be impractical. To close this Chapter, some key recommendations specific to nuclear reductions and disarmament are synthesized below.

The underlying theme is nuclear "de-emphasis," that is, revamping Cold War procedures and reducing stockpiles. The nine specific steps that succinctly highlight recommended gradual actions for weapon-states call for de-targeting aim-points of missiles, de-alerting missile-launch procedures, de-mating warheads from missiles, de-MIRVing re-entry vehicles, de-creasing nuclear arsenals, de-fending against ballistic missiles, de-weaponizing outer space, de-militarizing fissile materials, and de-minishing nuclear infrastructures. Those are the most explicit actionable items. It is not too soon to get some or all of the nine nuclear de-emphasizing steps under way.

In order to promote deep cuts in nuclear arsenals, some strands of helpful strategic considerations are extracted for the time-prioritized list in the following page. It suggests actions arranged according to perceived threats to national security.

Mutualism and Pragmatism. Benefits and difficulties associated with different approaches to national security need to be openly recognized, especially those once viewed through a prism that differentiated dissimilar superpower cultures. Russia has taken on an outlook that stresses international stability, whereas the United States has, since World War II ended, emphasized the importance of nuclear dominance.

If one had to choose just one key phrase to summarize this book's overall suggestions, that phrase would be *preventive* deterrence: maintained by nuclear and non-nuclear forces, gradually and guardedly reduced as warranted by "reciprocal measures." Preventive deterrence might very well include carefully delimited missile defenses. In order to maximize the value of preventive deterrence in a highly interactive world, multinational "mutualism" needs to be invoked: A nation's sense of security depends on both confidence and ability to defend itself against likely threats; therefore, its perceived defensive (as opposed to offensive) posture depends heavily on assured international stability. Mutualism reflects recognition of common interests and joint endeavors.

[First Things First].
Recommended nuclear-policy goals presented in a somewhat time-prioritized order:
- restrict targeting patterns for long-range missiles
- place constraints on missile-launch alert status
- proceed with RV warhead-reductions
- cut warhead arsenals to lower levels
- engage in multilateral negotiations on missile-defense and outer-space
- make reductions in warheads and fissile-materials
- continue conversion of nuclear arsenals and infrastructure

Mutualism does not mean comprehensive submission to an external authority; it means voluntary but frequent cooperation in regional and international affairs. An equivalent goal is "multilateralism" — the practical manifestation of alliances drawn up to fit the contemporary situation. Multilateralism is a practicable alternative to the unilateralism promoted by the G.W. Bush administration. Manifested in the form of NATO and the WTO, multilateralism had a major role in moderating the Cold War.

A studied view of mutualism is that

[We] may have to plot a course that flows from the premise that the United States is now inextricably part of an interdependent community of nations that will have to rely on each other to satisfy their respective interests and goals.... Mutualism views regional rather than global structures as the foundation of the emerging international system; it maintains that international cooperation is more likely to occur when states exercise responsibility for solving their own problems rather than when solutions are hierarchically imposed by overarching political structures and institutions.

William Pfaff, columnist for the *International Herald Tribune*, warned that a clash between Europe and America that might arise out of industrial competition in the globalized and deregulated international economy. He viewed "hegemonic pronouncements" from Bush-administration Washington as a revival of "manifest destiny," leading to an inevitable, destabilizing and dangerous competition with Europe. The antidote is improved international balance and extended peaceful accommodation.

Returning to the more specific job of dealing with Cold War legacies, "nuclear pragmatism" is shorthand for an approach that advises "quasi-static equilibrium" — stepwise nuclear reductions punctuated by pauses of stabilized consolidation. As each stage of de-emphasis is satisfactorily achieved and experienced, confidence can build to proceed further. Although the risk of a nuclear blowout has already lessened considerably, the weapons and their mechanisms for delivery remain armed and ready; that's the main driving force for gradual, rather than abrupt de-escalation.

While this book focuses on nuclear arms control, especially between former antagonists, the growing threat of terrorism — especially through weapons of indiscriminate casualty — is not minimized. Terrorism poses a qualitatively different danger, largely unrelated to residual nuclear arsenals. Terrorists have their own structures, adherents, weapons, and recourses. However, their coercive capability would be exacerbated by access to poorly safeguarded weapons that amplify injury.

Aside from these preceding limitations, an informed, less-panic-prone perspective has been placed herein on risks associated with the atomic age.

We — the four prime contributors to this book — don't think of ourselves as ideological alarmists, but rather informed realists. We think it will take several decades, possibly as long as the half-century bilateral Cold War engagement, to reverse fully the dangers and consequences of excessive nuclear buildup. In the meantime, global stability is highly dependent on "nuclear mutualism," the understanding that nuclear and non-nuclear nations must help each other avoid disaster from accident, malign intent, or out-of-control nuclear escalation.

Early in this book, we identified a reflexive Cold War process that suggested the image of "nuclear shadowboxing." Now, a weary and more coexistent world community is gradually (but fitfully) refraining from boxing against its own nuclear shadow. Thus, the burden of dangerous and unnecessary weapon and material inventories can be eased, enabling orderly withdrawal from the nuclear precipice.

****

Coordinated, time-bound nuclear reductions logically must follow a series of distinct steps, each intended to ease nuclear tension by trimming shared hazards to national security and human vitality.

Rather than going "cold-turkey" or acting precipitously, or unnecessarily risking national and mutual security risks, a program of gradual de-emphasis is advised. Some measures are functional and some are tangible.

Various constraints on missile launching have been proposed because long-range, land-based ballistic missiles have short flight times and cannot be recalled (unlike bombers) and are vulnerable to a first strike (unlike submarines). An early goal would be to verifiably reduce imminent danger from long-range missiles by a series of constrictive measures: de-targeting, de-alerting, and de-mating. These functional reductions can progress from those that would be qualitative and procedural to those that would be quantitative, physical, and verifiable.

De-targeting and de-alerting are comparatively simple steps to diminish risks that no longer prevail since the demise of the Soviet Union.

Warhead de-mating is the next rational step because it would physically separate warheads from their delivery systems. This is a tangible tension-relaxing measure that reduces the risk of possible hasty, unauthorized, accidental, or misinterpreted ballistic-missile launches. Warhead de-mating would institute demonstrable barriers and insert irreducible delays. Warhead de-mating is easily verifiable.

The next logical major challenge in drawing back from the nuclear precipice is to implement a systematic transition to low inventories of warheads. Economic conditions and pressures also give a boost to this palpable approach.

De-MIRVing is one of the best and most perceptible ways to reduce the number of deployed warheads and to improve nuclear stability: It reduces the missiles' attractiveness as targets, while alleviating the impact of precipitous retaliation. (If the off-loaded weapons are disassembled, it will expedite the tempo for "de-creasing" arsenals.)

Intrusive verification is key to mutual confidence in a negotiated nuclear stand-down. Stability can stem from knowledge of contemporary conditions— an awareness maintained by national intelligence-gathering and enhanced by cooperative measures.

As the number of offensively positioned strategic missiles and warheads diminishes, the feasibility and value of a ballistic-missile defense become more reasonable. In fact, a negotiated defense initiative (ballistic-missile de-fending) would preserve the current arms-control network—yet improve the prospects of viable defense because of its cooperative nature.

Protecting orbiting surveillance satellites that have numerous defensive and stabilizing attributes leads to the inclusion of space de-weaponizing as an important adjunct to de-fending.

In order to induce the largest possible delay in reconstitution of nuclear arsenals, existing warheads need to be progressively dismantled and destroyed, with their nuclear materials disposed of in an irreversible manner (fissile de-militarizing). Eventually the nuclear complexes that created these arsenals would have to be reoriented by revamping (de-minishing) their mission so they no longer are dependent on the production of atomic weapons.

Thus, "de-emphasizing" Cold War weaponry in (nine specified) parallel and sequential steps briefly summarizes with sound-bite simplification this culminating Chapter/Volume on incremental, functional, gradual, and tangible nuclear reductions.

DENOUEMENT

So much complex information and opinion have been presented to the reader that the need exists to distill some key thoughts into this final book segment. While each Section, Chapter, and Volume has had its own summary, this denouement contains selected highlights and conclusions from all three volumes of Nuclear Insights.

Our collaborative and comprehensive book undertaking started in 1993, soon after the Cold War ended. In the course of time, relevant information from current events has been incorporated, and updates have been added. Much of the same text and organization found in Nuclear Shadowboxing *has been retained, thus allowing traceability of sources, references, and supporting details omitted from* Nuclear Insights.

Four policy-seasoned nuclear scientist/engineers contributed to the two volumes of Nuclear Shadowboxing, *while* Nuclear Insights *is a modified abridgement offered up by the lead author.*

Volume 1, subtitled Cold War Weaponry (An Insider History), *is the informative, corrective, and interpretive history of events from a unique vantage point of involved nuclear scientists and engineers. Volume 1's intensive historical evaluation reflects comprehensive technical insight. Its broad policy evaluation is consistent with personal experience shared in common with members of government and the public. Volume 1 is uniquely "informative" because of our comprehensive treatment as insiders. It is "corrective" is the result of careful and knowledgeable research and involvement. That Volume 1 is "interpretive" simply acknowledges our own "take" on events. No matter what, at best our insider history merely complements other Cold War expositions.*

Volume 2, Nuclear Threats and Expectations (A Knowledgeable Assessment) *concentrates on accumulated contemporary threats and challenges resulting from the Cold War. Because of our across-the-board technical experience, we can present a comprehensive and knowledgeable technology assessment. At the same time, our analytical and experimental tendencies are reflected in what we think is an objective analysis of the nuclear heritage. Our children and their children will indeed inherit "a mingled yarn, good and ill together."*

This Volume 3, Nuclear Reductions (A Technically Informed Perspective), *concludes the trilogy, addressing a constructive and highly specific theme: how and what to implement regarding changes in nuclear policy, strategy, and stockpiles. For the nuclear arsenals inherited or inspired by World War II and the Cold War, here are key (rhetoric) questions: What will reduce untoward risks, and what are the downsides in reducing their size and hazards? While we indeed lean toward massive reductions, the reader should be aware of our nonconformity with authority and our impatience with subjectivity.*

One more caveat: we respect statistical and scientific humility, especially when it comes to our own analysis and prescriptions.

The Clash of Scorpions

Nuclear weapons — the "ultimate," decisive means for military conquest or perceived defense — were devised during World War II and used to terminate hostilities with Japan. Even before that finale, the seeds of confrontation—however parsed between Communism and Capitalism, between East and West, between the United States and the Soviet Union — were to grow into nearly a half-century of contentious Cold War.

The analogy of two scorpions in a bottle (without implying moral equivalency for the adversaries) comes to mind. From 1945 through 1991, the United States and the Soviet Union struggled against each other until the latter's collapse. That conflictual period is referred to as the "Cold War" because the two countries never actually fought each other directly with weapons as they would have in a "hot war." Instead, the hostile nations engaged in political and psychological posturing, carried out surrogate wars, and stockpiled conventional and nuclear weapons—all in an effort to outgun each other in the event that war did break out between them.

The action-reaction cycle characterizing the arms race was actually driven more by "domestic" issues on both sides rather than by real threats, "communist" or "imperialist." The perilous competition materialized in the form of nuclear shadowboxing, as the predecessor book was titled; each side was goaded more by its internal politics, economics, military strategy, culture, and ideology than by the actions of its perceived adversary. Figuratively, in sparring with ever-increasing nuclear feints, each side was often boxing against itself.

That is not to say there weren't real threats and dangers to Western nations from the Soviet monolith. Having "been there, seen that," the coauthors of *Nuclear Shadowboxing/Nuclear Insights* understand the reality and temper of the times. One of us served with the American military during the Korean War, one with the Soviet Navy, one helped make Soviet nuclear weapons, and the other strived for international arms control. We all experienced the fears and hopes of the extended struggle. We do not minimize the suffering of many who were bridled by the Soviet yoke, nor do we agree with the bellicosity and brinkmanship of the governments.[*]

Even regional clashes at remote battle lines were enactments of proxies assigned by the United States and the Soviet Union as a stand-in for the ideological bipolar competition. Go-betweens did the dirty and dangerous work.

In any event, according to hardliners, "specialists" — even Nobel-prize-winning scientists — were supposed to avoid meddling in "political courses of action." They were (and still are) advised to defer to politicians and political appointees the decisions that put nations or civilization at survival risk.

---

[*] Although we generally have neutered the terminology, the pursuit of an imperious and dangerous policy to the limits of recklessness — brinkmanship — was highly male-dominated.

Learning Experiences

But, the question persists, did our communal institutions gain any fungible and lasting understanding from this traumatic period where concepts of humanity nearly vanished?

Some American Cold Warriors rejoice in "victory" having "won" without the weapon of last resort being used. We are less sanguine, finding that the Soviet Union broke apart not so much because of "containment," "resolve," or "military power," but *despite* the compulsive gamesmanship and brinkmanship that characterized excesses of the time. Having been close to Armageddon is hardly something for warhawks to brag about.

The costly U.S. hard-line approach for Vietnam failed to achieve success through four U.S. presidential administrations. Refusal to realize implications resulting from the colonial era's end was one reason; another was the "domino" theory's fallacy that nations would sequentially fall to communist ideology unless there was military intervention. The Cuban missile crisis, a prime example of brinkmanship, ended as a "success" story only in the sense that it didn't flare up into a nuclear exchange.

More than one "true-believer," in government office when the superpowers were moving toward the abyss, has taken autobiographical comfort in the opinion that President Reagan's policies put an end to the Cold War. That's highly disputable; after all, the Soviet Union didn't dissemble until 10 years after Reagan became President!

Historians concur that, long before Reagan was in office (in the 1980s), the USSR was already declining—economically, politically, and culturally. Perhaps Reagan accelerated the demise, but less confrontational policies might very well have made it happen at much less cost and risk to the West — and, who knows, maybe even sooner.

Here's another lesson of importance: National security is closely related to economic health. Nations now depend less on armies, territory, or natural resources and more on the ability to adapt and integrate into the global economy—a transition that ideological communism and the arthritic Politburo were unable to accommodate. Governance needs to engage a mutually reinforcing dynamic in economic and security relations with other nations and the world community.

In any event, American and Soviet political leaders during the period deserve at least some political kudos for avoiding a nuclear clash.

Yet, both sides sustained confrontationalism by feeding half-truths and speculation to their constituencies, and, worse, they believed their own palaver. Both sides, dwelling on fear created by hardliners, sanctified bellicosity with little attention to restraint and arms control. The public purse was a convenient and deep reservoir.

Worst-case analysis almost caused our mutual undoing. With each military or technological advance inducing the other side to follow suit, tit-for-tat response was embodied in the development cycle for nuclear systems — fission weapons, thermonuclear weapons, ICBM, MRV, MIRV, ABM, SLCM, MX, etc. This action-reaction cycle ratcheted the participants to dangerous encounters and outrageous budgetary excesses.

A preponderance of nuclear weapons did not help nuclear states prevail in regional conflicts like Berlin, Korea, Vietnam, and Afghanistan. Nor did the atomic bombing of Japan and the subsequent four-year U.S. nuclear monopoly prevent the Soviet Union from taking over Mongolia and Eastern Europe.

Pervasive myths were propagated about arms control, treaty compliance, and military strength. At the time, the myths became detached from accessible realities; their delusory role was not confirmed until the lengthy struggle terminated. Some fantasies are still promoted.

Three specific and dramatic claims in the 1950s through the 1980s of "gaps" in military strength and "windows" of vulnerability gave rise to a series of U.S. political decisions that upped the arms-race stakes with more or better nuclear weapons. We know now that these claims were overstated in order to further the goals of hardliners. If anyone was "ahead," it was the United States. The forces of deterrence were at such high levels that additional weapons added little to national security but increased the risk of mutual nuclear annihilation.

The Soviet Union engaged in wholesale infringement of freedom and boundaries in the Eurasian land mass. These were atrocious examples of human-rights abridgement. Corresponding American trespasses of sovereignty took place in South and Central America, the callous anti-communist government abductions in South America, and the Western-supported mercenary forces in Africa.

The Mi Lai massacre in Vietnam by U.S. troops helped turn public opinion against the American endeavor to prop up an anti-communist regime. While House recordings disclosed the initial Nixon-administration attempt to suppress photos of the massacre. World opinion was as indignant then as it has been since the recent wars in Afghanistan and Iraq after photos and reports surfaced depicting American abuses of detainees in their custody. Although the Cold War corpse was only recently buried, the ghost of lessons-never-learned is being resurrected.

Other already-forgotten lessons could be drawn from the efforts of post-Cold-War governments to sacrifice human life and rights in their pursuit of perceived national security. Outrageous post-millennium policies still divert attention from long-existing violations in distant parts of the world and encourage others to violate human or prisoner rights.

That important lessons have not rooted themselves deeply is a conclusion easily discerned in the aftermath of President G.W. Bush's 2003 invasion of Iraq. A year after the unsanctioned invasion, the *New York Times* admitted that it had been hoodwinked into accepting unconfirmed reports of WMD and Al Qaeda in Iraq. The newspaper's belated explanation was that their reporters and editors were suckered by a "hall of mirrors," where one creative image was reflected many times to make it appear like a cascade of truth. (Physicists have an apt analogy: Lasers rely on mirrored surfaces to reflect and amplify stimulated light emission.) To us, the hall of mirrors recalls the image of shadowboxing.

Pentagon neocons, not surprisingly, were among those that helped amplify the Iraq threat image. Many were former Reagan-administration conservatives or hardliners who resurfaced in the G.W. Bush administration to carry on their unconsummated agenda.

Cold Warriors tend to be dismissive about public manifestations of dissent. Worst than that, the war hawks enabled or condoned the suppression of constitutional rights in order to sustain nuclear pugnacity. Dissident individuals, groups, and organizations had to maintain vigilance and withstand pressure against such excesses of the time. While public dissent failed to stop massive nuclear buildups and confrontations, think what might have happened without such prickling and prodding!

Some legacies can be considered to be "good," some "bad," some "tolerable." Some might have occurred anyway, perhaps at a different time. In terms of future peace and stability, it is important to acknowledge the Cold Warrior's lemming-like parade toward a deep chasm.

Even World War I was unexpected, unpredicted, and lacked a substantial *casus bellum*. In 1914, alliance leaders and strategists assumed that it would be no more than a localized, low-intensity conflict. Those moderating forecasts and scenarios were proven wrong when the tragic war propagated throughout Europe and elsewhere.

It's dismaying that these admonitions might not be sufficiently and widely understood enough to preclude recurrence of near-fatal nuclear conflict — by design, carelessness, error, or accident. Some indelible traits of past virtual hostilities continue, especially dependence on nuclear umbrellas and "tripwire" deployments. The new millennium was greeted by NATO with nuclear-weapons assigned to its bases in Europe. Forward-based American troops have long been positioned in locations, as in South Korea, where nuclear temptation might come up if overrun by an invasion. Prominent individuals hold on to the dogma that nuclear weapons remain useful components of national policy.

Hopefully, this trilogy contributes to clarification of the frightening and self-fulfilling consequences of perilous nuclear buildups and to realization that their reoccurrence is not unimaginable.

Unthinkable

United States weaponry was qualitatively light years ahead of the Soviet Union. That advantage did not prevent a stalemate, nor did it assure dominance. The inability to prevail was frustrating to the hardliners, though it did not stop them from putting into place procedures and weapons that might have enacted the "unthinkable."

Thanks to bragging rights, we can now read memoirs of proud American and Soviet Cold Warriors who rationalize the huge weaponry buildups. Their own writings betray the ascendency of fear. Even with 30 years of domination in warheads and delivery systems, U.S. insiders beat the drums of anxiety. Little of this would have passed open scrutiny had the real arsenals and their consequences been publically disclosed. Eventually the Soviet Union caught up and passed the United States in some of the statistics, but never did that lead to any of the feared dominance; the nuclear standoff continued to the point of exhaustion.

In 1962, after both superpowers had deployed nuclear-armed ballistic missiles outside their own national boundaries, the Cuban crisis was one of the first to escalate to the threshold.

Such brinkmanship did not occur spontaneously; paranoia over national security was often pervasive and tended to dominate the public and intra-government debate — if open discussion was allowed, because much of the discussion that took place was kept secret. National-security paranoia was incorporated into the "worst-case analysis" used to justify expansion of nuclear arsenals. Whether through ignorance or error, extreme fears assumed about the adversary were usually imaginary.

Much of this can be attributable to the nuclear priesthood, a cabal of hardliners that knew how to gain leverage through the use of an imbedded, self-perpetuated, and expanded military-industrial-laboratory-congressional quadrangle of influence. The "priesthood" that maintained militant momentum was a faction of career officers, business entrepreneurs, influential scientists, optimistic technologists, operations analysts, and ambitious policy makers.

Individuals in the United States who promoted brinkmanship were often self-styled defense intellectuals; they banded together in "think tanks" that reinforced primal fears of national vulnerability to the perceived threat of communism. To the antithetical side, Western imperialism was the monster.

Those precarious conditions resulted from a process of assessment and planning that was war-oriented, with little historical precedent and few checks or bounds. War gamers processed nuclear warfare as though it were a programmed contest between reasoned and informed adversaries, both of whom understood and agreed to play by the rules. In war gaming, escalation control was presumed, despite possible failure of communication links and despite nearly instantaneous strategic-weapon delivery and unprecedented firepower. Illusions about the utility of nuclear war and expectations for a favorable outcome were widespread. There was little understanding of statistical uncertainty or catastrophe theory.

The intractability of President Reagan and his staff can be traced to advisors who were mostly very conservative Republicans (and some converted Democrats): Richard Perle, Edward Teller, Paul Nitze, Richard Pipes, Jeane Kirkpatrick, Fred Iklé, and Frank Gaffney. The inclusive Committee for the Present Danger ingratiated itself into the government, almost swallowing the executive branch whole.

Devastating weaponry in the hands of unpredictable leaders and nations — which had pre-emptive, escalatory, or delegated policies — risked all civilization as hostage for nearly a half century. The durability of societal institutions and survival of civilization as we know it was wagered in a form of high-stakes, "no-limit" poker — although the final round of the contest ended with more tractable "pot-limit" stakes. (For those of you that don't play poker, it was all or nothing for a while — only one winner — until mutual restraint in wager size became prudent.)

Fortunately, the extended period of marginal conflict reached closure with a soft landing for the ideological collective that was the Soviet Union, rather than a more violent unlimited convulsion.

Cost of the Cold War

The economic cost of this half century mired in quasi-military engagements was severe for both superpowers.

The CIA found that the Soviet collapse had more to do with their own domestic economic and social problems than with their inability to keep up with the U.S. military increases. Long before Reagan's ascendency, the CIA was aware that the Soviet Union since the 1970s was on track for economic and political implosion. American government leaders had been routinely briefed on these findings.

In the United States, long periods of recession or inflation correlated with periods of heightened conflict (like the war in Vietnam and the Reagan buildup).

Certainly the financial cost was substantial all around. The United States alone charged the public purse about $19 trillion, of which almost $6 trillion —nearly a third — went to nuclear arms. The Soviet Union became essentially bankrupt. Allied nations ended up spending huge parts of their GNP. Subjugated or war-burdened nations came out impoverished.

To be balanced against cost, both financial and societal, are less-quantitative "benefits" of the Cold War: The West was able to withstand any further incursion of communism and Soviet tyranny (although Soviet hegemony prolonged its power over people and lands for many decades).

Another appropriate question comes up: Given that the West "won" and that "freedom" was protected for those outside the iron curtain, what was the role of nuclear weapons in bringing this about?

Actually, *nuclear weapons won nothing during the Cold War*. In fact, they have never really "won" anything! We repeat, nuclear weapons have *won* nothing, not a war, not the peace.

Even Japan, for all practical purposes, was defeated before the atomic bombs hastened its surrender.

But before Japan surrendered, and despite the brandishing of nuclear weapons by the United States, Stalin took over Manchuria and retained Eastern Europe. That was when the United States had a supposedly imposing atomic monopoly. The Cold War got off to a bad start for the West.

True, further Soviet expansionism was stymied by the deterrence effect of nuclear weapons. In particular, some would argue that nuclear weapons kept Stalin at bay immediately after World War II.

Not liberated by nuclear weapons, Eastern Europe remained and suffered under the Communist yoke for more than 40 years.

None of the hallmark proxy wars — in Berlin, Korea, Vietnam, Cuba, Afghanistan — was decided by nuclear weapons. The Berlin airlift took place in 1948 through 1949 when the United States still had a monopoly in nuclear weapons.

The Korean War ended in a stalemate (still divided by the military demarcation line and demilitarized zone after half a century). Vietnam was eventually taken over by the indigenous Viet Cong.

The Carribean/Cuban confrontation, an edgy situation where nuclear weapons were flouted, settled down to a draw. It ended with the Soviet Union recalling its nuclear weapons from Cuba, and the United States withdrawing them from Turkey.

The Soviets lost in Afghanistan; the West later won, but without using nuclear weapons.

In fact, historian Gregg Herken specifically notes that "One reason for military leaders' reluctance to recommend use of the bomb in Korea was its obvious unsuitability for that kind of war. [There were] no useful targets in Korea for the bomb." In Herken's book, *The Winning Weapon: The Atomic Bomb in the Cold War 1945-1950*, he concludes that the atomic bomb was "then plainly a wasting asset," even though Truman "entertained the notion that the atomic bomb could still be America's 'winning weapon' as late as spring 1952...."

Herken defined the "deadly illusion [of security]" to be

the fallacious assumption concerning the utility of nuclear weapons: their supposed efficaciousness in diplomacy and their alleged capacity to avert military confrontations. It survives as myth to the present day [1980].

The analysis here in *Nuclear Insights* extends Herken's interpretation to the present day. Nuclear weapons have not been the "winning weapon" that many strategists and policymakers made them out to be.

Although the definition of a hot "war" has sometimes been stretched beyond classical meaning, its past consequences have included systematic and organized violence, physical casualties, and the shedding of blood. Certainly in this sense, nuclear weapons won no war. While significantly influencing the "Cold" War, nuclear weapons were brandished but not used. Nuclear arsenals that were much smaller might have had the same deterrent outcome.

The past ineffectiveness of large nuclear arsenals, rather than simple military deterrence to consummate aggressive policy goals, casts doubt on any possible future role for them by existing weapon states. Even so, "have-not" nations and leaders tempted to develop nuclear weapons recognize their value as deterrence against infringement of sovereignty. Meanwhile, transformation of the Cold War threat to that of global terrorism has given rise to suicidal militants who would not be deterred by nuclear weapons; even worse, they might not hesitate to use acquired nuclear explosives to advance their cause.

Loyal Opposition

How effective was the public at-large in reducing Cold War confrontational immoderation? Judging by the great lengths hardliners and apologists have gone to ignore, downplay, or contest assertions of the public's effectiveness, that alone would suggest the loyal opposition was more than merely unimpressive and ineffectual.

Solitary private individuals rarely impacted the Cold War truculence, but— when networked with others and with public-interest organizations — conscientious people did have a significant role in moderating excesses. Although one can't claim that public pressure brought the superpowers to their knees, public involvement did temper some excessive buildups and threats. Previously classified tapes and records are proving how much public opposition was feared by government officials. Public-interest organizations were effective in consolidating and focusing its opposition.

While public influence was quite visible in the West, it was relatively inconspicuous in the Soviet Union, though a few "refuseniks" and high-level "apparatchiks" openly resisted military excesses.

Not many military programs were killed, although some—like missile defense, nuclear testing, ASAT, MX, and Euromissiles—were trimmed to less-threatening proportions. Public intervention in the West was successful at least in slowing the quest for unlimited killpower and moderating the deployment of untimely weapon systems.

All of this occurred despite the heritage of Stalinism and McCarthyism, when every possible government tool was invoked to instill fear in individuals and groups who spoke out or opposed national-security policies.

The United States Constitution was not enough to protect law-abiding citizens from being smeared by association. Those who exercised their First-Amendment rights often found themselves the target of investigation. Whenever protesting against excessiveness in the nuclear-arms race, personal and institutional safeguards were put at risk, while overt promotion of national security was cloaked in patriotism.

Contrary to abiding convictions of "true believers," President Reagan's flip-flop about the realities of the Cold War was largely due to peace-movement pressure. He and his senior advisors belatedly began to accept and promote the concept of offsetting asymmetries — the idea that one category of weapons could offset another. The Reagan administration eventually went to the bargaining table with a more conciliatory stance. Saber-rattling slowly subsided and negotiated arms reduction followed the pathway promoted by peace activists and other moderates.

Testimonials aside, no convincing evidence exists that Reagan's highly touted "bargaining chip" approach to negotiations had a substantive role in a world characterized by asymmetric response. Star Wars was of little value to the United States as a bargaining chip because the Soviets easily and much less expensively thwarted it by installing countermeasures and by increasing offensive forces.

Were it not for the scientists, NGOs, and foundations that coalesced to bring independent information, analysis, and pressure to bear on the complex nuclear issues, civilization might have moved closer and more frequently to the brink of nuclear disaster.

Ultimately modernized communications technology enabled the propagation of protest, helping to undermine the totalitarian Soviet state. Dire post-Hiroshima predictions of catastrophe did not come to pass. Although the dangers of nuclear war were obvious to both East and West, public involvement had to maintain pressure in order to lessen the risk.

Contentious Arms-Control Issues

Hardly any buildup — in contrast to build-down — of nuclear arms met with universal public approval. As the public became less deferential to government decisions, the balancing of national security, budget deficits, and unnecessary risk became a source of contentious debate in the West.

Both governments often strived to avoid, to delay, to drag out, or to water down arms agreements. Some would use their bully pulpit to convince uninformed constituencies of the negative aspects of arms accords, without providing balanced portrayals. Propaganda and disinformation prevailed, particularly in the Soviet sphere.

Some credit for the non-radioactive course of events probably is owed to the existence and growth of international institutions, such as the United Nations. Neutral institutional structures provided an important, perhaps only, venue for input from nations not directly in the line of fire, as well as a forum for discussion by adversaries who sometimes were not talking directly to each other.

What We're Stuck With

Derived from 20th-century world wars is universal hope that beneficial lessons will prevail in the new millennium, and that we are not doomed to repeating the mistakes, practices, or risks that culminated in an almost mutually destructive face off.

Difficulties bequeathed are still around: radiation contamination, institutional disruption, dilapidated nuclear security, and huge nuclear arsenals.

In the past century the most devastating, potentially self-destructive, weapons of humankind were created. At the same time beneficial technologies emerged, leading to global economic reconstruction, environment improvement, and trade expansion. Relationships between nations have gradually improved (though tormented by a surge of transnational terrorism). Democratic changes toward openness (the people's voice) have spread to more nations, despite the harrowing past and rocky road ahead.

Outside of vast military holdings, another heritage is the vast and complex legitimate international business involving nuclear fuel at various stages of isotopic enrichment and in diverse chemical, metallurgical, and radioactive forms. It is difficult for non-technical individuals to understand its complexities, and correspondingly perplexing to technically trained people trying to grasp its full magnitude.

Although national, regional, and international institutions have evolved to manage and control the worldwide nuclear enterprise, it is a difficult task. Fortunately, only certain high-grade nuclear materials are amenable to weaponization, and those materials and their means of production, storage, and transportation have additional security restrictions.

Fissile-material production management has always been recognized as a potential Achilles heel of international nuclear commerce. Plutonium can be controlled in a comparatively straightforward manner because is normally only accessible in highly contaminated form, not usable for weaponization. Enriched uranium, under universal safeguards too, is somewhat easier to weaponize, so a high degree of international visibility is normally maintained on its production.

A contentious situation has evolved from Iran's domestic production of low-enriched uranium justified as a reliable, indigenous, and economic source of fuel for its nuclear reactors. Although the NPT protects the rights of nations to produce low-enriched-grade uranium, Iran has hedged on allowing full-scale IAEA inspections intended to ensure that enrichment to military grade is not taking place.

Suggestions have been made to establish an international nuclear-fuel bank that would sell suitably enriched uranium to Iran (and other nations lacking indigenous production facilities). However, many issues would need to be resolved regarding equal business opportunity, hard-currency sales, national-autonomy, and energy independence. Western nations have had a near-monopoly on this profitable enterprise. Base on the prevailing capitalistic business-model approach, Iran would be justified to enrich and sell reactor-grade uranium on the world market.

Ubiquitous Radiation

Nuclear-production processes were accompanied by radioactive byproducts expeditiously stored or dumped, awaiting future remediation. The most substantial harm caused to workers and the populace resulted from exposure to dangerous levels of toxic chemicals rather than radioactivity.

"National-security," as a justification, was often misused to study radiation effects without gaining consent. That practice was consistent with autonomies and immunities assumed by governments and with personal disassociation that officials relied on to enlist people as subjects of experiments.

Several reactor accidents produced widespread anxiety. None of the accidents, it has turned out, caused the off-site harm predicted in the newsmedia, which broadcast and amplified the fears of radiophobic and misinformed individuals and organizations. While radiation in the immediate vicinity of accidents was sometimes lethal or otherwise harmful, its danger diminished rapidly with distance.

Low daily levels of radiation are a natural occurrence, well tolerated, perhaps even beneficial to the vitality of life. Current radiation-protection standards are overly cautious, and therefore unnecessarily expensive; the standards should be based on practical thresholds. Radiation protection is slowly being reformulated to recognize a practical threshold below which net radiogenic damage to health cannot be detected or might not occur.

Because of excessive caution embodied in current radiation-exposure standards, they are unnecessarily expensive to implement; they should be based on practical thresholds. The overly conservative LNT linear-extrapolation hypothesis for low radiation dosages is increasingly being seen as invalid. At low doses, it is more of a "worst-case" approach, a methodology used during the Cold War to scare people and politicians into spending more on wasteful and dangerous arsenals.

Institutional Legacies

For individuals who suffered and survived Soviet tyranny, it no doubt left a tormented and lasting imprint. For those outside the iron curtain, the threat of mutual assured annihilation was incessant. While subjugated peoples broke free to engage their own destiny, the risk of nuclear annihilation remained for the West. That continuing, though now greatly abated risk, tops the list of Cold War institutional legacies.

Government secrecy, a problematic legacy, is not the sole or dominant answer to assuring national security. Although nuclear-information control has delayed other nations from devising their own weapons, it has not kept them out of the nuclear club. Conversely, secrecy over nuclear-utilization policy has ended up sustaining risky strategic adventures detrimental to the unwary public.

This book is intended to do its part in discouraging nuclear nations from retaining, or other nations from contemplating, self-destructive nuclear arsenals and expensive arms-races. Governments are implored to turn their attention to redressing the last half-century of domestic social neglect.

Although the two *Nuclear Shadowboxing* volumes satisfied our inquisitiveness and sense of fulfillment, we mainly wanted to systematically review and share what went wrong (and right). That's why we embarked on its 12-year publication time-devouring endeavor. Now, to help translate and interpret those results for a wider audience, *Nuclear Insights* is herewith presented.

Very pervasive has been the aura of protection and secrecy conscripted by governments in the name of national security. This was an outgrowth of secrecy tenured long enough to protect incumbents from the consequences of public disclosure. Governmental executives have found ways to suppress information for generations and lifetimes, long enough — but for a few audacious whistleblowers — that future historians will have to delve deeply into redacted archives to find buried truths.

Unsubstantiated Fears About Low Radiation Doses. The case for excess cancers coming from natural radiation background is so weak that it serves better as an example of difficult epidemiology and unsettled science. A long-term international study about villagers who lived downstream from the Mayak weapons complex in the southern Urals found that few died from possible radiation-induced cancers. Many of these individuals inhaled tobacco smoke, lived in poverty, or imbibed alcohol — three life-risk factors that were prevalent in the Soviet Union at the time and are much more strongly associated with various forms of cancer.

A survey of more than 400,000 nuclear-power-plant workers in 15 countries has been interpreted to suggest that between 1 and 2 percent of the deaths may have been due to radiation. But a serious shortcoming in that study has been flagged: Smoking may account for a large share of deaths mistakenly attributed to radiation.

There are three basic weaknesses for these two cited studies: (1) inaccuracy in determining radiation doses for the particular individuals that died of cancer, (2) poor statistics because of the small number of deaths attributable to radiation, and (3) inability to sort out life-risk factors correlated with known carcinogens.

Many places on earth have background radiation 50, 100, or more times higher than the sea-level annual average. In these places epidemiological studies were started, and they produced a remarkably consistent and benign picture: People there have either the same or a slightly lower chance of cancer compared to their less-irradiated compatriots.

Radiation exposures comparable to or less than natural background should not be considered a health hazard. This conclusion is reinforced by a number of carefully controlled epidemiological studies which have examined the effects of small added doses of natural, occupational, and medical radiation.

Progressive Aftermath From Reactor Accidents. A 20-year aftermath international assessment of the Chernobyl-reactor accident tells us that its mortality consequence had to be adjusted — downward from earlier estimates of tens or even hundreds of thousands future cancer deaths, lowered to 4000 *at most*. Since the ill-fated 4000 statistically estimated persons were still alive after 20 years, it would be more correct to use terminology such as "life shortening" or "reduction in life expectancy," rather than "final death toll" for the calculated projections.

Even more important is that most of the calculated radiation-induced cancer deaths will probably never occur. No doubt many people will develop cancer, but it will have nothing to do with the radiation they received from the power plant. It has turned out that the thousands of highly exposed liquidators — the firemen and emergency workers at Chernobyl— have experienced no more cancer than the average Russian population.

Twenty years after the 1986 Chernobyl reactor accident, the adult death toll was 56 persons, 47 of them being emergency workers.

Out of thousands of children who have since died of thyroid cancer (largely due to chronically inadequate health care in the Soviet Union and lack of iodization of kitchen salt as in the West), statistically perhaps a dozen of them might have contracted it as a result of radiation exposure from Chernobyl.

Other than thyroid cancer, there was no evidence of any increase in cancer or leukemia rates among local residents, nor was there evidence of decreased fertility or of a higher rate of congenital malformations.

By and large, *not found* were profound negative health impacts to the rest of the population in surrounding areas, nor widespread contamination that would continue to pose a substantial threat to human health.

The *mental health* impact of Chernobyl has been the largest public problem unleashed by the accident. For the 350,000 people moved out of contaminated areas, relocation was a "deeply traumatic experience" which often left them unemployed and homeless.

The other major reactor accident was in 1979 at the Three Mile Island facility in Pennsylvania (The accident occurred in the TMI-2 reactor; TMI-1 is still operating today). Although no injuries or adverse health effects resulted, it revealed much about risks resulting from a reactor melt-down. Over subsequent years, nuclear industry and government regulators have made numerous adjustments and changes to foster a more vigilant safety and reliability culture.

The reactor core was destroyed. A couple of days later some radioactive gas escaped outside the site boundary, but not enough to cause any exposure above normal background levels to local residents.

No injuries or adverse health effects resulted from that accident except for immediate, short-lived mental distress. In fact, since initial 1957 operation of the Shippingport civilian nuclear power plant, not one single fatality has occurred as a result of the release of radiation from a commercial nuclear power plant in the United States.

Although epidemiological studies found no increased incidence of cancer or other physical illness as a result of the TMI accident, they did find evidence of psychological stress.

The accident has done much to shape the safety culture of today's nuclear-power industry. Before, few, if any, in the nuclear industry would have believed that reactor operators would fail to recognize the symptoms of a stuck-open relief valve. Even fewer would have believed that operators would reduce water flow in the face of symptoms of inadequate coolant.

Three decades after the worst nuclear accident in U.S. history took place in Pennsylvania, renewed growth of nuclear energy has eclipsed the event.

In the annals of human endeavor, few human-engineered complex structural systems have such a lengthy record with operationally safe, functionally efficient, and environmentally benign experience.

Nuclear Security in Russia

After the Soviet Union fell apart, one of the more serious national-security concerns for all nations was whether Russia would be able to safely and securely manage its inherited nuclear stockpile.

Centuries-old traditions of disrespecting the law affect the young Russian nation. Adding disruption and corruption — nuclear, economic, governmental, and social — is not a healthy beginning. Russia's estate includes huge integrated military-civilian complexes, many of which remain nearly inaccessible.

Russia, as successor, has Soviet nuclear weapons, power reactors, propulsion reactors, and kindred infrastructure. To reduce the risks of illicit nuclear trafficking and to strengthen safeguards against diversion of materials, international programs of security support have had to be funded and implemented.

President Putin and his successors have evidently been trying to restore Russia's superpower status. That's given rise to an independent strategy concerning nuclear-based defense and international collaboration. As a result, one must conclude that, on balance, Russia's nuclear drawback has been in remission lately.

It is in the ultimate security interest of other nations to help Russia improve its control of fissile materials and facilities; broaden the transparency of past production, disposition, and current stocks; end production of weapons-grade nuclear materials; dispose of the excess weapons and materials; and increase incentives for nuclear workers to adopt civilian-oriented jobs.

Nuclear-Armed Nations

The end of the Cold War did not necessarily mean a drawdown of excessive nuclear armaments nor the demise of dangerous nuclear-use policies. Too many bureaucrats and vested interests, both East and West, remain to rationalize the past, abhorring cold-turkey withdrawal.

Nuclear arsenals had about 60,000 weapons at their peak (out of a total production of 100,000). Upwards to 50,000 are still stockpiled. Besides reducing quantities, tasks for the new millennium include dealing with deployed warheads and delivery systems; reducing risks of accident, error, or override; and taming the tendency of the military-industrial-laboratory-legislative complex to preserve the inherited nuclear complex.

In the seven-year period from 1977 through 1984, more than 20,000 U.S. nuclear alerts were triggered — all false indications that ballistic missiles were attacking America. We don't have comparable estimates of how many times Russian missiles were armed for massive retaliation against apparent missile assaults from the West.

Nuclear arsenals are being created by nations such as Pakistan, India. and North Korea; and other nations have moved closer to joining the less-exclusive club.

Here we are, 60 years after Hiroshima and Nagasaki, with bestial nuclear arsenals. The fear that nuclear weapons will again be used — in anger, defense, mistake, or accident — has engendered decades of sociological and political trauma. Nations would benefit by national security that depends much less on a war footing, however surreal the idea might seem now. Negotiated arms control and its overt verification would be helpful in such a transition.

Potential Nuclear Terrorism

The potential for malign distortions of nuclear technology by sub-national groups (terrorists) or individuals is not without justified concern, and protecting or countering such actions or attempts is difficult and emotionally charged.

In actuality, planned or attempted nuclear terrorism has been limited to a few documented situations: allegations, warnings, and chance events. Some cases of poisoning with radioactive substances have been publicized, although other poisonous substances would have been quicker or more lethal.

Little doubt exists that vulnerabilities exist or could be found that would lead to the spread of radioactivity to an unprepared public. Examples include sabotage of a nuclear reactor, theft of nuclear fuel or waste, or even acquisition of a nuclear weapon. The prospects of acquiring and detonating a nuclear weapon have been far overblown. The possibility of fashioning a so-called "dirty bomb" is small, and the consequences inconsequential as far as potential physical harm.

A deliberate, disguised nuclear attack would be extremely difficult to pull off. Far more likely would be a highly counterproductive nuclear threat by a pariah nation. More effective is simply nation-state development of one or a few nuclear weapons in order to ward off perceived enemies and to gain negotiating leverage. The actual implementation of a nuclear threat is entirely counterproductive, as improbable but achievable it might be.

[Can Terrorists Build Nuclear Weapons?]. Eminent nuclear-weapon-laboratory specialists have noted many inherent technical and logistical obstacles to be faced by terrorists before they could construct nuclear explosives. Formidable barriers must be overcome to succeed in acquisition of high-grade materials from storage sites or nuclear transport [Emphasis added].

A terrorist group would ... have to proceed deliberately and with caution to have a good chance of avoiding any mishap in handling the material, while at the same time proceeding with all possible speed to reduce their chance of detection.

The time factor enters the picture in a quite different way. In the event of timely detection of a theft of a significant amount of fissile material — whether well suited for use in an explosive device or not — all relevant branches of a country's security forces would immediately mount an intensive response.

The production of sophisticated devices ... should *not* be considered ... a possible activity for a fly-by-night terrorist group. In summary, the main concern with respect to terrorists should be focused on those in a position to build, and bring with them, their own devices, as well as on those able to steal an operable weapon.

In order to combat potential nuclear terrorism, national and international agencies should have missions, capabilities, and resources to prevent nuclear or radiological attack or detonation, and be able to respond and manage the consequences. This would include capabilities for nuclear search, device deactivation, recovery operations, field assistance, aerial and ground detection, consequence management, and medical support. Nuclear forensics is a viable science that can assist in early detection and damage mitigation.

Although there are over 100 research reactors around the world, many becoming tempting targets for terrorists, their theoretical vulnerability provides little real risk for public danger.

Internal (domestic) nuclear security should be reinforced in nuclear-weapon states, especially Russia and other nations that have unstable constituencies or volatile neighbors. Nuclear materials and weapon stockpiles need to be internally and externally safeguarded; this could be enhanced by international policies, supportive organizations, and viable response mechanisms.

It's no exaggeration that reductions in nuclear-weapon stockpiles and weapons-grade materials would throttle potential nuclear terrorism. Scaling down the magnitude of nuclear arsenals and supplies would simultaneously minimize potential sub-national and accidental nuclear threats.

Beneficial Impacts of Nuclear Technology

Non-military nuclear technology has turned out to have benign, productive applications: clean power generation, long-range naval propulsion, nearly limitless energy supplies, consumable-resource prolongation, outer-space adaptability, and advancements for modern living. Nuclear technology has proven absolutely essential to life-saving diagnostics, medical treatment, health, and sanitation.

Nuclear technology has contributed to the quality of human life through promotion of peace, prosperity, energy, well-being, and a better future. Of increasing importance to the growing world population, is atomic/nuclear energy, which has none of the palpable drawbacks of fossil-energy sources. Atomic energy is non-polluting; it emits no greenhouse gases; it is the safest source of large-scale energy (Chernobyl notwithstanding), and it is essentially inexhaustible. If implemented wisely on an international basis, base-load atomic power will not contribute to the proliferation of nuclear weapons — and in fact might well reduce the motivation by helping to remove supply shortfalls in energy, water, and fuel as sources of international tension.

[Nuclear Power Role in Conversion of Military Stockpiles]. Precisely 439 reactors worldwide were providing 372 GWe in 2007, and 48 new plants were under construction to reach 410 GWe by 2015. Significantly more reactors were in planning stages to overcome those being retired.

All of this has increased fissile-fuel supply requirements, which are largely based on mining of uranium ore, which itself depends more on the spot price offered than on reserves. Consistent with the current slowdown in the growth rate for nuclear power, there has been a corresponding reduction in the amount of natural uranium being mined.

Although uranium-supply self-sufficiency doesn't exist now, it not a long-range nuclear-power resource limitation: More mining is like to happen based on higher price incentives, and the use of thorium would make up for any future shortfall. In addition, fast-fission could considerably increase effective fission-energy extraction.

An interim gap in fissile-material supply can be and is satisfied with secondary resources. These are (1) conversion of civilian and military stocks of uranium and plutonium accumulated during the Cold War, including dismantled nuclear warheads, (2) recycling of MOX, a mixture of U235 and plutonium extracted from fuel rods that have undergone a partial burnup/depletion stage in nuclear-power reactors, and (3) re-enrichment of uranium tails.

As with all technologies, public understanding, competent management, and ceaseless control are essential.

Troubles and Perils Still

The Cold War's benign closure has been universally applauded. Its heritage of formalized arms control and transparency must be considered a positive outcome of that fractious period. Yet, despite the ensuing era of relative peace and international stability among the nuclear powers, still needed are mutual vigilance and assistance.

In 1995, the liftoff and stage separation of a scientific rocket was detected and initially erroneously construed by the Russian military to be a U.S. submarine-launched missile. Even though several years had gone by since nominal Cold War closure, President Boris Yeltsin reportedly (and reluctantly) had to activate his "nuclear keys" for the first time while awaiting threat confirmation. In a matter of minutes, a momentous human decision had to be made on the basis of (flawed) technical input and (archaic) operational procedures. Because of mistaken radar blips, the arming stage of all-out thermonuclear retaliation was spasmodically initiated, though eventually nullified by human reasoning.

Presidents Bush and Putin signed an agreement in May of 2002 to remove some strategic weapons from deployment, adding little assurance that headway had been made in actually reducing the risk of hair-trigger nuclear response to false alert. The situation has been aggravated by an increasing number of other nations armed with ballistic missiles.

After aggregate nuclear arsenals peaked in 1985, the number under deployment soon dropped by a factor of three. The remaining warheads were stored or later refurbished. Since the Cold War's demise, the warhead-reduction process has tapered off.

Enough fissile material was produced to make a total of 100,000 weapons; most of this fissile source material remains in existence. And, we must ask rhetorically about targets: Now that the Cold War stress is over, are there still tens-of-thousands of credible targets for 50,000 nuclear warheads?

Perils are inevitable if nuclear materials and their production processes that might be inadequately unsafeguarded. High on most lists is the danger that additional nations will acquire nuclear weapons. Another menace is that non-state entities (chiefly terrorists) might acquire nuclear (or other) methods for inflicting mass casualties. The unrest in and around Pakistan is worrisome.

Deterrence arguably kept the fractious Cold War nuclear confrontation from drifting into outright war; yet, the Berlin blockade and the Cuban missile crisis are just two reminders that the superpowers approached the doomsday brink. The fact that deterrence worked is not so much in question as is the Cold War oversupply of nuclear weapons: Heedless of necessity and expense, stockpiles reached grotesque proportions.

Huge arsenals are still retained by the two former geopolitical adversaries. Strategic weapons and delivery systems remain poised to attack or counterattack. Tactical nuclear weapons have simply been reshuffled, for the most part, behind national borders. Defense policies and grand strategies of both superpowers still lean heavily on nuclear forces.

As for ballistic-missile defense, the only real hope is engagement in a cooperative multinational program designed for interception of a few missiles. A defense-protected drawdown would favor mutual reductions in offensive missiles and, conversely, nuclear reductions significantly enhance the feasibility of ballistic defense. The modified (theater) missile-defense system proposed by the Obama administration has many elements of the cooperative or mutual missile-defense system advocated in this book.

"Benign" nuclear earth-penetrators are — one might say — another pipedream; they are designed expressly for war-fighting as would be the development of new, small battlefield nukes. The Pentagon is sending the world a message that the United States is determined to pursue nuclear dominance indefinitely — that nuclear arms restraint, according to unrepentant warhawks, is already an outmoded relic.

To match the Pentagon, the geopolitical circumstances in Europe and Asia, and its budget, Russian contemporary military doctrine relies heavily on nuclear weaponry. India's and Pakistan's increase in size and effectiveness of their nuclear arsenals complicates unilateral, bilateral, and multilateral arms reductions.

The Proliferation of Deadly Hazards

Civilized nations have increasingly recognized risk from proliferation of non-conventional weapons: nuclear, chemical, and biological. The chem/bio menace has grown because of increased transnational terrorism. To deal effectively with this varied threat spectrum, a consistent and universal nonproliferation policy is needed; also required are acknowledgment and progress in accommodating the desire of various indigenous peoples for recognition, self-determination, equality of economic opportunity, improvements in human rights, and respect for the Earth's environment.

While nuclear weapons are indeed capable of mass destruction, chem/bio are better described as weapons of indiscriminate casualty. Although terrorists routinely inflict harm and instill fear with readily accessible conventional explosives, some perpetrators have realized that the threat or use of non-conventional weapons amplifies the intended dread and cost of recovery. Traditional law-enforcement and paramilitary responses are insufficient by themselves. Supply interdiction, public education, response preparedness, and reduced incentives are needed to ameliorate — but also could not eliminate — these threats. Coming back to roost are many policies of foreign-alliance and weapons-supply that once provided materials, knowledge, and productive capabilities to then-friendly nations.

Many nations have offered to cooperate in outlawing all forms of biological and chemical weapons. The United States is yet to explicitly disavow its interest in offensive bioweapons, nor has it renounced non-lethal chemical weapons, nor allowed transparency within its own military and commercial facilities.

Myths have propagated about terrorism — its origins, impact, and cures — and about use of radiation to advance nefarious goals. So-called dirty bombs that disperse radioactivity, another overrated threat that has gained mythical proportions, would create psychological and economic impairment but little physical harm from radiation.

Returning attention to true weapons of mass destruction, nuclear proliferation unchallenged endangers not only the weapon states, but also international norms to cease and desist from heavy-handed posturing or outright warfighting.

In early stages of nuclear-weapons development, the United States selectively ignored or even actively assisted the spread of nuclear weapons to friendly nations.

Weapon-state parties are still not fulfilling Article VI provisions of the NPT, the most accepted treaty in international relations: It calls for steps toward nuclear disarmament. Testing of nuclear weapons has been outlawed by the CTBT, which is slow coming into force as a universal and verifiable treaty because of U.S. failure to ratify. It has been difficult to gain a consensus on respect and enforcement of international nuclear norms, especially when the United States abstains itself or practices a dual standard. Nonproliferation leadership and example would have to be demonstrated by the nuclear-weapons states in order to successfully face down new applicants.

It should thus not be incomprehensible that nations have hedged by covertly developing nuclear materials and capabilities, at least to the threshold of weaponization, and North Korea to its nuclear-explosive demonstrations.

Through historical and technical analysis, this book has provided testimony that nuclear weapons are not for war-fighting; they are much better left unused as instruments of deterrence, as they were during the Cold War, when much smaller arsenals would have served the same dissuasive purpose. That successful deterrent role is a constructive present-day lesson for have-not nations, some of which have been pursuing their own nuclear capability when faced with axis-of-evil threats to their sovereignty. Moreover, global terrorism has emerged as a pressing danger, ineffectualizing nuclear weapons which do not deter terrorists, but which, conversely, remain attractive to suicidal militants who might not hesitate to detonate nuclear explosives.

Residual stocks of biological and chemical weapons and techniques for disseminating them add another ugly threat to the peaceful coexistence of rival interests. The emergence of a more vehement and suicidal form of transnational terrorism adds stress on institutional controls of nuclear, chemical, and biological substances that could be exploited indiscriminately.

Terrorism in its many potential unconventional forms — nuclear weapons of mass destruction or biological, chemical, and radiological weapons of indiscriminate casualty — has become a discomforting, potentially lethal, post-Cold-War phenomenon. Stringent government controls, along with suitable disincentives, can help contain possible harm and damage.

A particular setback to nonproliferation of all types of weapons was the 2003 invasion of Iraq, touted largely as an attack against an Al Qaeda haven for terrorism and against a maverick state possessing or capable of manufacturing chemical, biological, and nuclear weapons. None of these allegations were authentic.

Improving Security

Have we cleaned up the Cold War mess? Not really! Physical remnants include nuclear weapons, fissile materials, strategic delivery systems, and radioactive contamination. Institutional vestiges include heavy-handed foreign policies, information-suppression practices, and outmoded military strategies.

Besides necessitating a prolonged agenda to help Russia with its burden of weaponry, other military legacies are haunting peace and stability in the new millennium.

Nuclear-weapon states outside the original five pose special difficulties for regional and international stability, and more nations, or maybe terrorist groups, might develop or obtain nuclear weapons.

Regarding the huge physical and virtual inventories of fissile material, multilateral agreement has been emerging that nuclear-reactor burnup is the best method for demilitarization. To degrade weapons plutonium and uranium, consumption in commercial nuclear reactors is the most proven and satisfactory institutional means because the reactors despoil fissile materials while producing heat and electricity to offset the cost of demilitarization. Vitrification, an alternative that leads to chemical immobilization and underground burial of fissile materials, has too many technical, regulatory, and custodial difficulties to warrant its added expense.

[Demystifying Depleted Uranium]. A residue from natural uranium fissile-isotope separation, depleted uranium has positive or negative value as commodity or waste. It has rendered both civilian and military functions. Depleted uranium's intrinsic value depends on marketability.

It has been exploited in nuclear weapons and nuclear reactors, as well as commercially in radiography shielding, coloring agents, and aircraft trim weights. Existing and future stockpiles of depleted uranium have significant value as feedstock for re-enrichment or as fuel for plutonium-breeding nuclear reactors.

Because of its high density (over 19 g/cm^3), depleted uranium metal has been utilized by the military as armor plate or in armor-piercing projectiles. Neither of these particular functions make use of its intrinsic fissionable nuclear attributes, which have considerable potential as a source of nearly limitless energy.

For a given mass, depleted uranium metal has a smaller diameter than an equivalent projectile made of elemental lead, thus less aerodynamic drag and deeper penetration due to a higher pressure at point of impact. After piercing armored-tank plate, some depleted uranium residue might be dispersed into the environment as metallic or oxide fragments, but it is very unlikely to become a powder that could cause an inhalation hazard.

Because of its very low radioactivity, battlefield depleted uranium does not represent a radiological threat; that is, it is not a viable candidate for use in a "dirty bomb" or radiological-dispersion device. Any cancers or deformities found in the battlefield would be much more likely caused by other toxic agents, not depleted uranium.

Natural uranium is ubiquitous in trace amounts; it is found in soil and water almost anywhere.

Over a million tons of depleted uranium remain in national and multilateral stockpiles. Its harmful properties derive from high density — rather than mild chemical reactivity or low-level radioactivity — so battlefield uranium merits no more, no less proscription than any other conventional means of warfare.

So far, only states that were not full-scale signatories of the NPT have produced nuclear weapons. Iraq, a party to the treaty, was far from success when stopped by outside intervention; North Korea formally withdrew from its treaty obligations;

Iran has been hedging about inspections that might reveal too much. South Africa signed the NPT only after it ridding itself of a small indigenous nuclear arsenal. India, Pakistan, and Israel never did sign up.

Misplaced worries about peaceful uses of atomic energy and civilian plutonium had significantly delayed progress in arms control, disarmament, and nonproliferation.

Nuclear Zeal

Tending to get in the way of balanced nuclear management in the new millennium is an eventful history of excessive public and official zeal regarding atomic weapons and technology. Going back through these three Volumes, extracted below some examples of what might be considered as dogmatic or callously enthusiastic adherence or opposition to some nuclear policies and practices. Some of this admittedly is Monday-morning quarterbacking; moreover, the happenings listed are skewed because many Cold War decisions were made as well as possible under the circumstances and with the available information and resources.

Strategic errors were made, in retrospect, both in the East and the West; nuclear zeal and lack of objectivity were exhibited on both sides of the iron curtain. And, of course, some of these practices or occurrences ended long before the Cold War was over, while some continue even now.

Even though *Nuclear Insights* has been mostly list-phobic, here's an exception where specifics are compiled and consolidated in one place. The reader would have to refer to the supporting text for explanation and background.

In order to highlight different strands of excessive, risky, or unwarranted zealousness, the particulars are divided into three nuclear-related categories — weapons, technology, and radiation. For those who want to avoid the lists, or simply derive their gist, the three categories of overzealousness are: (1) dangerous or callous ventures involving nuclear weaponry, (2) misuses and abuses of nuclear technology, and (3) subjective alarmism.

Dangerous or Callous Ventures Involving Nuclear Weaponry. These are examples that represent excessive enthusiasm or unmitigated zeal for political policy that depended on nuclear explosives:

► Needless or premature use of nuclear weapons to hasten the capitulation of Japan
► Excessive, chronic buildup of nuclear arsenals
► Overt or implicit threats to use nuclear weapons to achieve policy objectives other than deterrence
► Questionable government decisions to develop staged thermonuclear weapons
► Governmental promotion of radiation-fallout shelters
► Nuclear brinkmanship, especially in the Caribbean/Cuban crisis
► Implementation of strategic policies that favored unstable nuclear deterrence
► Excessive development and deployment of tactical nuclear weapons
► Forward deployment of tactical and intermediate-range nuclear weapons
► Limitless test explosions of nuclear weapons on the surface and in the atmosphere
► Testing peaceful nuclear explosives without evaluating environmental and public impact
► Military development of nuclear reactors to be flown by aircraft
► Unconscionable claims made for nuclear ballistic-missile defense

- Planning and testing for radiological warfare
- Excessive secrecy about nuclear-radiation releases into the atmosphere and environment
- Politically motivated allegations of arms-control violations
- Unjustified dependence on worst-case analysis for military strategies and decisions
- Unfounded claims about bomber, missile, and vulnerability gaps
- Research and development of counterproductive anti-satellite capabilities
- Developing weapon systems as arms-control bargaining chips
- Insufficient resources for preventing accidental nuclear detonations
- Downplaying the importance of public dissent regarding nuclear explosives
- Inexcusably allowing excessive occupational exposures to radiation and toxic substances
- Putting expedience ahead of health and the environment when discharging and dumping radioactive wastes
- The MX western-shuttle deployment scheme
- SDI's flamboyant promotion of unproven space-based weaponry
- The development and deployment of neutron bombs
- Allowing the military-industrial-scientific-congressional nuclear-promotion quadrangle to get out of hand
- Abandoning decommissioned nuclear ships and submarines in shallow-sea graveyards
- Blacklisting and denying security clearances based on unfounded or irrelevant allegations
- Overuse of nuclear secrecy
- Insufficient attention to methodical nuclear espionage
- Bombing and sabotaging foreign nuclear reactors
- Invading Iraq under the guise of WMD possession
- Funding and testing atmospheric electromagnetic-pulse (EMP) nuclear explosions despite excessive collateral effects

Misuses and Abuses of Nuclear Technology. Here's a compilation of overly zealous or thoughtless applications involving peaceful nuclear technology:

- Excessive nuclear-reactor buildup without adequate domestic and international safeguards
- Industry promotion of nuclear-fuel recycling without suitable proliferation protection
- Inadequate reactor-safety design, engineering, and operation for Soviet-built graphic-moderated, water-cooled power reactors
- Insufficient attention to the export and control of sensitive nuclear technology, especially uranium enrichment
- Inconsiderate disturbance of local water ecology for reactor cooling
- Overly hyped projections of nuclear power
- Serious underestimation of institutional difficulties and reactions in introducing new technologies
- Unnecessary occupational safety and environment risks allowed
- Inadequate inventory accountability and custody of radiation sources
- Shortchanging accountability and safeguards for nuclear materials
- Carrying out human radiation experiments without informed consent
- Continued adherence to the sham of nuclear science born secret
- Government abuse of its own secrecy regulations and laws

Alarmism. Equally vexing and counterproductive is ideological or reflexive opposition to beneficial nuclear roles and functions. Lack of objectivity—that is, the absence of systematic and comprehensive assessment — is usually characterized by a preoccupation that lacks systematic scientific methodology. For example, unbridled environmentalism is counterproductive, especially when it runs against a healthier, safer, and more benign environment and systematic ecology.

Alarmists: It takes one to know one. Having being involved in drawing public attention to concerns related to nuclear weaponry, the principal author of this book and his contributing colleagues have been alarmists about nuclear excesses. One or all of us have taken issue with government policies regarding fallout shelters, nuclear testing, ABM deployment, military-spending priorities, the military-industrial-laboratory complex, insufficient institutional arms control/verification, MIRV deployment, inadequate safeguards, neutron bombs, proliferation incentives, MX deployment, and excessive government secrecy – to recall some described in Volume 1.

We have at one time or another had to disagree with fellow scientists and environmentalists about some specific technical issues: radiation effects, reactor accidents, nuclear-waste storage, reactor-grade plutonium, dirty bombs, and proliferation risks — considering ourselves to be less alarmist, more studied in our approach to these complexities (see Volume 2). A knowledgeable perspective has resulted in us being lukewarm to the fisban, open-minded about communal missile defense, sanguine about proliferation management, cheerleaders for deep nuclear reductions, but merely temperate bystanders regarding nuclear abolition.

Here are some specific results of excessive alarmism to be found in these three volumes of *Nuclear Insights*:

▸ An unnecessary 15-year or more delay in upgrade of Siberian nuclear dual-purpose power and heat reactors
▸ Obdurate opposition to shipping and storage of spent nuclear fuel at Yucca Mountain without providing viable alternatives
▸ Chronic delays induced in storage of military wastes at the secure WISP facility
▸ Compulsive opposition to nuclear power disguised as concern for safety, proliferation, or waste-storage
▸ Resistance to a fissile-production ban because reactor-degraded plutonium is not included
▸ Opposing weapons-plutonium demilitarization by MOX burnup
▸ Tacit support to atmospheric gaseous and particulate pollution by burning coal compared to clean nuclear power
▸ Blindly supporting depletion of fossil resources and emission of carbon into the atmosphere
▸ Overreaction to Chernobyl accident and overstatement of its consequences
▸ Excessive reaction to TMI accident and gross overstatement of its health impact
▸ Promoting discreditable vitrification for storage of weapons plutonium
▸ Inciting and exploiting public radiophobia
▸ Exaggerating the potential impact of "dirty" bombs
▸ Resistance to benign public applications of radiation
▸ Overemphasis on horizontal rather than vertical proliferation
▸ Religious advocacy of the baseless linear (LNT) radiation hypothesis

Particularly annoying is a narrow academic alarmist approach, a galling example being anti-nuclear power stridency (see box, next page).

Another disturbing practice is over-hyped preoccupation about terrorism, illustrated graphically and simplistically by depicting the potential impact of nuclear explosions in American cities. (A number of technical and logistic factors militate against successful acquisition, transportation, emplacement, and detonation by terrorists of an achievable nuclear explosive: The contributors to this book have personally or collectively visited or worked in many nuclear facilities — including those involved in national, regional and international safeguards — witnessing the secure protection of nuclear materials and weapons, even from complicit insider access. We couldn't say impossible—being well aware of statistical capriciousness —nor do we promote a lessening of safeguards, or underestimate the potential for insider threats, but it would be extremely difficult for terrorists to acquire sufficient usable nuclear-explosive material or full-up weapons, especially without timely discovery of the plot.)

"Dirty bombs" (crude radiological-dispersion devices) are another pedantic dream/nightmare. They belong to the fiction and drama side of entertainment.

Unbounded nuclear proliferation and the propensity for nuclear war are representations created by more by imaginative than realistic assessment. We too have exploited some such fears, but the fact remains, after more than a half century of peaceful passage, that nuclear reductions should be carried out for more tangible, cautionary, and mutual motives.

[Anti-Nuclear-Power Stridency]. For many years, especially during the Cold War, peaceful applications of nuclear power came under fire from compulsive opponents. Nonetheless, as the benefits of nuclear power became ingrained worldwide and the weapon risks subsided, anti-nuclear zealotry has noticeably diminished.

Most objections to nuclear power were based on somewhat abstract considerations that have not materialized. For example, nuclear-weapons proliferation has been limited to far fewer nations than feared, and nuclear enrichment and reprocessing risks have been suitably managed and contained. All of this success and progress is contrary to every alarmist prediction.

Much opposition, primarily academic in nature, was based on abstractionist arguments that avoided comparing specific benefits in relation to realistic alternatives. Often ignored were government-subsidized "externalities" of pollution caused by fossil-fueled power sources. (Coal-burning has explicit health and environmental costs — such as chronic mining fatalities, acid-rain stimulation, and mercury emission — that are implicitly subsidized, despite very real and harmful liabilities.) Fossil-fuel oxidation unavoidably results in copious carbon-dioxide emission.

By putting forth freewheeling still-unproven assertions of low-level radiation-induced cancer, the abstractionists have disproportionately influenced reactor siting, radiation-waste disposal policy, and nuclear liability. They take advantage of the confounding complexity of nuclear issues.

A half-century of experience has proven nuclear power to be one of the safest, most reliable, and economical emission-free technologies created by human beings. Reactor accidents, though quite expensive and troublesome, have been infrequent and of minimal public-health consequence. (These remarks are not intended to discourage open-minded public vigilance.)

While a few anti-nuclear partisans still cite potential nuclear "threats," such as terrorism and "dirty" bombs, their exaggerated fears overly stress hypothetical possibilities rather than empirical evidence.

Moreover, nuclear reactors provide the only economic and viable means of demilitarizing weapons-grade fissile materials, thus providing irreversible closure for nuclear reductions.

Having once been certified alarmists, we now speak more assuredly about the benign aspects of nuclear power, about moderate radiation exposures, and about the role of nuclear reactors in destroying fissile weapon materials. We fear malevolent exaggeration and stridency more than we fear controlled technology.

Having assembled this testament, we feel entitled to the designation "environmentalist" just as much as anyone who respects, conserves, improves, and avoids harming human and natural surroundings.

We have acquired a distaste for an all-too-frequent lack of objective standards in dealing with contemporary technology, especially the nuclear totality. The most anti-scientific aspect of environmental enthusiasm is often a lack of statistical humility, revealed by failure to recognize or assign measures of uncertainty to statements, declarations, or conclusions.

What is Shaping the Future?

The 2002 Moscow agreement by Presidents Bush and Putin, which removed some strategic weapons from deployment, did not directly affect existing inventories of nuclear warheads; moreover, strategic reductions via the START II treaty and START III goals were tossed into the dust bin. No pronouncements since then gave reason to believe that significant multilateral nuclear arms-control and non-proliferation progress would have been made during the G.W. Bush administration.

With the former ideological conflict now history, both the United States and Russia are self-deterred from attack on each other; in fact, far fewer warheads and delivery systems would do the job. No nation poses a strategic threat to the two superpowers.

Although still glamorous and dangerous, massive nuclear arsenals are obsolescent. Designing or making more nuclear weapons is not beneficial to national and international security, nor is the testing that accompanies their development, nor the proliferation and resentment they stir up.

Despite the increased aura of peace and security, U.S. military spending under Bush was approaching the level that existed when East-West tensions were at their peak. American warhawks have trying to settle old scores (i.e., the "axis of evil") rather than focusing on tractable threats to national security.

Because the military budget, estimated by the administration for two presidential terms, added up to the jolting value of $3 trillion — more than half of what was spent on the entire U.S. anti-Soviet cause ($5.5 trillion) — additional stress was placed on an already sagging U.S. economy.

Many sacrifices had already been made by both superpowers. The time has come for recovery, partly by attending to the hopelessness, corruption, repression, and poverty that spawned civil distress in hot spots around the world. Both Russia and the United States have retained some communist-era practices of detaining suspects without judicial review, and continue to practice some heavy-handed forms of surveillance and secrecy. Military aid is being sent to nations run by dictators.

Indeed, mutual assured destruction remains — years into the 21st century. The risk of self-induced annihilation seems to have become second nature to nuclear-armed nations: Despite comforting assurances by government officials, few political dangers have diminished, while more hazards from nuclear weaponry persist.

Strategic policies and residual arsenals need to be moderated in such a way as to bolster national security while providing public relief. For example, declaratory policies could be implemented that would immediately decrease at least the perception of potential precipitous or unauthorized nuclear attacks. This could be followed by deliberate and systematic reductions in the most threatening strategic armaments. Such voluntary or negotiated reductions would not at all endanger the security of nuclear-weapon states.

Citizen pressure continues to be the main bastion of effective resistance to ill-conceived, ideologically motivated programs against developing new nuclear-weapon systems, prematurely deploying ballistic-missile defense, and retaining surplus nuclear arsenals. Outsider opposition remained important as the G.W. Bush administration, favoring unilateralism, governed in a manner unhurried, indifferent, or resistant to many constructive lessons from the past.

With the passage of time, other threats to world stability have come to the fore. These include — in no deliberate order — internal ethnic or boundary disputes, terrorism, hunger, illicit trafficking, disease epidemics, substance abuse, environmental degradation, population pressure, social inequities, political tyranny, economic disparities, violent crime, supply shortcomings, and a diverse multitude of "isms."

Unfortunately, the competing agendas of well-meaning environmentalists have sometimes impeded progress in nuclear reductions.

Treaties and Negotiations

During the Cold War, a framework of agreements, mostly codified in national and international law, was structured for nuclear-arms limitations — to keep nations from engaging in dangerous and expensive attempts at one-sided imbalances of power. Success of this arms-control framework depended on a number of factors, including the creation of verifiable conditions. Good faith also entered into the composite, but self-interest remained the strongest motivator, so as to avoid deliberate, accidental, or unauthorized nuclear incidents and attacks.

The G.W. Bush-administration's rejection of the universal-negotiation paradigm and of international-affiliation advantages came at a bad time. While the existing arsenals of the United States and Russia have been largely aimed toward each other, Russia — more than ever — has had great interest and incentive to reduce strategic and tactical nuclear inventories. Reductions would greatly and immediately improve each nation's respective national security, as well as conform to international norms for progress in nonproliferation and disarmament. The Non-Proliferation Treaty has been floundering as some "have-nots" have done more than simply envy the "haves."

A new multilateral strategic-reduction treaty should spell out comprehensive nuclear-arms stabilization procedures. That would provide a stable basis for constructive cooperation: drawing down excess nuclear materials and warheads, improving potential missile defense, dealing with terrorism, and stabilizing other critical geopolitical issues.

The success of international peacekeeping offers hope for the future, just as their limitations offer avoidance lessons. United Nations and regional peacekeepers have helped and legitimized the process of normalization and reconstruction. International agencies, having gained experience with treaty verification and monitoring, are in a position to carry out the mandates of new treaties to control or reduce armaments.

Lacking ratification by requisite nations (among them the United States), the CTBT global system for detection of nuclear testing has nevertheless been funded and implemented. The network had 270 monitoring facilities in 2009, aspiring for 340. Signals transmitted to the Vienna headquarters are sensitive to underground shock waves, underwater perturbations, low-frequency propagated sounds, and airborne radioactive byproducts — some or all indicative of underground, underwater, or atmospheric nuclear-explosions. Nuclear tests conducted underground by North Korea in 2006 and 2009 were detected "very well," according to reports. Included in the global network are Iran and Israel. North Korea, India, and Pakistan are yet to sign up.

President Obama pledged during his election campaign to move forward on many of the stalled arms-control and nonproliferation treaty negotiations, and early in his administration significant steps were taken to implement his pledge.

The Nuclear Genie

Taking everything into consideration, nuclear weapons have been and are still attractive and feasible for deterrence against nuclear or conventional assault by nations or coalitions: so much so that if nation-state security is not assured by multinational or international measures, nuclear proliferation is likely to continue. Alternatively, if intrusive international inspections, coupled with energy and security assurances, could be negotiated on a *quid-pro-quo* basis, nuclear proliferation to additional nations might be discouraged.

Pre-emptive attacks against nuclear facilities yield only short-term benefits. That's not the best way to stifle interest in atomic weaponry.

As for the residual Cold War nuclear states, the temptation will persist to use atomic weapons in response to some atrocity or simply to gain political advantage. It is an easy choice for domestic political gain. For example, the U.S. program to develop earth-penetrating nuclear weapons has addressed a narrow world view espoused by U.S. warhawks.

Recurrence of a stressful nuclear arms race is not an unfounded fear. Inhibiting its resurrection should, we think, be a central facet in stabilizing national and international security.

Nuclear planning, as exhibited by published strategic doctrine and declaratory policy, is not simply a matter national security. Since that is not in doubt for the nuclear-weapon states, domestic political stakes have overtly become the legitimating factor for nuclear forces. The former superpowers have openly justified retention and even expansion of versatile and operable nuclear arsenals in order to bolster their own self-centered interests. In fact, domestic politics might very well trump international political stability.

Knowledge and wherewithal for making nuclear explosives cannot be eradicated, but the *sine-qua-non* (the requisite ingredient for any nuclear weapon) is military-quality fissile material. Academic nay-sayers need to recalibrate their thinking in line with historical reality. To prevent treaty breakout or a new arms race, nuclear materials should be sequestered and source-material production needs to be banned. Going one step further, the 2000-ton legacy of highly refined weapon-grade fissile materials should be demilitarized, making it unsuitable for reconstitution of nuclear arsenals and too vitriolic for diversion by terrorists.

Maintenance of nuclear retaliatory capabilities, partly under the legitimate aegis of stockpile stewardship, has taken on a life of its own. In the United States, "maintenance" has become a code-word for "modernization," which is — as it was during the Reagan administration — a thinly veiled disguise for new designs and modifications that stretch beyond credulity. Stockpile stewardship certainly keeps the nuclear genie active.

As for latecomers, India only belatedly has revealed the scale of its full-scope nuclear-weaponization program initiated at least in the early 1970s. While this program was frequently accompanied by vigorous public denials, its recent fulfillment emphasizes that national governments have often gone t to great lengths to cover up and deny existence of nuclear-weapons and delivery-system development.

Taking into account the world community's proliferation experience (especially including Israel, North Korea, India, Pakistan, and South Africa), it's no surprise that denials by Iran about weaponization intentions are met with considerable skepticism (see box, next page).

All in all, though, nuclear weaponization seems to approaching a saturation point rather than being open-ended as once feared. The liabilities and disincentives of nuclear proliferation appear to be capping its perceived advantages and incentives.

If you ask whether it is possible to protect every bit of nuclear knowledge, stymie technological capacity, and sequester source materials, the answer is **no**: The genie cannot be stuffed in the bottle any more than all the toothpaste can be squeezed back in its tube. But if you are satisfied with more modest, practical goals, *yes*, compulsive threats posed by huge war-fighting nuclear arsenals could be thwarted if sincere efforts were made. And, *yes*, the proliferation of nuclear weapons to other nations can be forestalled.

Deep Cuts

Missile-defense proponents are fond of saying that one intercepted missile might save a million lives. Yet, the same projected benefit will more assuredly come from negotiated elimination of a single nuclear warhead through an arms-reduction pact. In fact, it would be a "twofer": At least two warheads would be demolished under agreed eliminations since reciprocity would be required. Imagine what demilitarization of a hundred or a thousand warheads would do (and there is no risk that missile-intercept will fail!).

[Nuclear-Proliferation Ambiguities]. Outrage against Iran's stonewalling about its nuclear program appears somewhat hypocritical. While the focus has been on Iran's alleged intent to weaponize, in another part of the Mid-East Israel appears to possess somewhere between 100-200 nuclear weapons. No one really knows because (1) Israel did not sign the NPT, (2) Israel has refused to confirm or deny nuclear weaponization, and (3) Israel possesses delivery systems qualifiable for nuclear weapons.

Iran, for the most part, occupies no one else's territories, while Israel — lacking a conclusive resolution — arguably occupies Palestinian territories. Although Iran has been very cagey about IAEA inspections, Israel has ignored all international inquiries into its nuclear capabilities. Conforming to and emulating worldwide tradition, Israel is using almost any means necessary to preserve its national entity and cultural identity.

Saddam Hussein's nuclear-weapons program and his invasion of Iran in 1980 provide precedent and concern to justify today's Iranian nuclear-weapon program. The Iranians have reasons to be suspicious of U.S. intent toward their country and their region. They have every right to seek respect, particularly after a history of abuse experienced at the hands of Western countries.

A real and scary nuclear standoff exists between India and Pakistan, neither of which are signatories to the NPT. In fact, the main concern has been not whether Pakistan will use nuclear weapons against India but rather whether the Pakistani security system can adequately protect the weapons from capture by terrorists.

Iran knows fully well that any attempt to use nuclear weapons against its neighbors, not to mention against Israel, would result in immediate retaliation. Meanwhile, the specter of a preemptive or retributive Israeli or U.S. military strike casts a shadow on Mid-East strategy and diplomacy.

Casting a broad shadow on all these machinations is the major nuclear-weapons states failure to systematically comply with Article V of the NPT.

Advocates of deep cuts believe that mutual and verified worldwide curtailment of warheads and fissile materials is an imperative for longer-lasting tranquility and security. To reduce the threat of nuclear war as an "inescapable backdrop" will require significant reductions in weaponry. Another imperative would be to have an effective international mechanism to moderate sources of friction that upset the balance of peace and power.

Russia, which inherited many thousands of Soviet nuclear weapons, has been mired in economic distress. Strategic nuclear-weapons of both superpowers remain on problematic alert, launchable by accident or mistake. Such unrelenting hazards have led some to counsel the complete abolition of nuclear weapons.

We cannot sign on for such a strong position. As George Kennan (quoted by R.C. Longworth) urged, "in a world of relative and unstable values [we must put away] the search for absolutes in world affairs: for absolute security, absolute amity, absolute harmony."

Nowadays, rather than absolute abolition, many arms-control/nonproliferation analysts have advocated the less-traumatic "deep cuts" in nuclear inventories. This is reasonable and consistent with *de- minimis* recommendations.

As a starting point, why wouldn't an arsenal of a few hundred nuclear warheads be sufficient to deter overt military attack? The superpowers now have tens of thousands: Why not reduce to thousands and then, if the sense of mutual security is favorable, trim to hundreds? Some optimists look forward to a residuum of a few or none.

The Non-Proliferation Treaty contains an unfulfilled obligation of the United States and other signatories to negotiate toward zero nuclear weapons and to permanently ban the testing of nuclear weapons. Yet, an enduring self-appointed nuclear-weapons priesthood has lobbied for arsenals at near-peak levels.

A well-intentioned atoms-for-peace program was initiated when the United States first had a nuclear monopoly; the program's evolution has remained inherently flawed because the international regime preserved nuclear-weapons states as a special class that never met its obligations to reduce armaments. Notwithstanding all NPT, IAEA, and voluntary efforts to curtail proliferation, peaceful applications will always be tainted by unfulfilled demilitarization.

Somewhat more imminent security concerns have dominated post-Cold-War world and national events, often displacing nuclear reductions to the back burner. During the G.W. Bush-administration, foreign and domestic policies deflected attention away from comprehensive nuclear reductions.

Contrary to misinformed interventionists, weapons-quality fissile source materials can be demilitarized in a manner that would be gainfully compensated by saleable power production. The more-dangerous nuclear materials first need to be corralled in order to reduce risk of terrorist acquisition. A "dirty" bomb to spread plutonium would be frightening, but only slightly more of a hazard than one that disperses any other radioactive (or other vile) substance.

Deep cuts in nuclear arsenals would have multiple benefits for all: reducing mutual risk, minimizing potential for proliferation, satisfying international agreements, accelerating the demilitarization of nuclear source materials, and improving the prospects for ballistic-missile defense.

Verification Issues

Much maligned by war hawks, treaty verification is an increasingly important mainstay of mutual arms control. Cold-war hard-liners opposed verification in order to fend off treaties and to avoid the perceived risk of giving away secrets.

Past verification experience is invaluable. Innumerable obstacles of the contentious times had to be overcome: substantial stakes (military and political) in the consequences, mutual and indelible suspicion, vested domestic interests, inadequate compliance technologies, public-relations difficulties, and persistent reluctance to accept the results.

A good model is START I bilateral verification. It was negotiated in a step-by-step fashion, codified with explicit directions, institutionalized with designated responsibilities and mechanisms, reinforced with treaty-implemented technology and procedures, backed up by national technical means and abiding diligence, and presented as a package to the public and to legislative ratifiers.

Verification of warhead dismantlement would be essential to obtain confidence about nuclear disarmament. The agreed process must be thorough, formal, and intrusive. The process would have to cover all rungs in a demilitarization ladder that includes removal of operational weapons from service, secure transport to holding sites, and storage prior to disassembly. At dismantlement sites, the chain-of-custody would have to be verified, as would be ultimate disassembly into fundamental subunits and constituents.

Verification can be negotiated so as to ensure confidence in arms reductions and avoid strategic handicaps from potential surprise treaty breakout. Multilateral verification has also shown to provide a significant boost in confidence and stability.

Of course, verification is a proxy term for an embracing process that includes on-site inspection and monitoring, national and international means of surveillance, intelligence collection and sharing, various intrusive and non-intrusive instruments, data declarations and information exchange, challenge and scheduled visits, tagging and sealing, public observations and reporting, and other confidence-building measures.

From the harrowing Cold War period has emerged comforting experience with the discipline of formal verification, proving itself to be a reliable and necessary process for enhancing mutual-security.

Negotiated Nuclear Disarmament

Near-term abolition of nuclear weapons is not a realistic expectation, but ever-downward levels of nuclear forces would reduce interim risk of self-immolation. Intrinsically dangerous to public health are the appearance and the reality of hair-trigger ballistic-missile alerts and loose command/control links.

During our careers, this book's contributors influenced, opposed, or witnessed first hand the growth of nuclear weapons. Now, together, we encourage a pragmatic and comprehensive stepwise program of nuclear reductions. Major cutbacks could be staged at a pace and scale that would not endanger international stability or national security.

To counterbalance regressive advice received from weapons developers and defense analysts, we offer viable alternatives. No longer justified is retention of chronic Cold-War-era suspicions and attitudes.

Only if deep cuts are conducted in a synchronized and equitable manner would nuclear-weapon states become serious about them.

If existing large stockpiles of plutonium (and uranium) were not demilitarized, it would leave risk and consequences of proliferation and terrorism. Some nuclear source materials already have natural protective radiation and isotopic barriers (as a result of reactor burnup); so they need not be further demilitarized. The primary goal is to ensure that plutonium has been degraded isotopically so that it cannot be put quickly back in weapons. Pending ultimate disposal or recycling in a sensible and economical way, nuclear material inventories must be safeguarded with high security and safety.

Our overall suggestion in a sound bite is "preventive deterrence." This is a moderate and pliable form of deterrence that can be sustained by a proportionate combination of non-nuclear and nuclear forces. The nuclear component would be gradually and guardedly reduced as warranted, coupled with other reciprocal measures, not the least of which might include agreed ballistic-missile defense. The concept of preventive deterrence evades the connotations of more aggressive alternative strategies to dissuade hostilities.

Preventive deterrence would be implemented by phasing out mutually shared nuclear risks. The nine key nuclear measures recommended for reversal of Cold War fixations and implements were condensed into a Chapter VII tabulation deliberately repeated in the box that follows.

[**Nuclear "De-Emphasis"**]. By making use of the Latin prefix *de-* (which means reversal or removal), and taking slight liberty with the English language, nine specific goals can be identified to reverse (or de-emphasize) the Cold War instruments of nuclear destruction:
1. "de-targeting" nuclear aim-points
2. "de-alerting" missile-launch systems
3. "de-mating" nuclear warhead s from delivery systems
4. "de-MIRVing" ballistic-missile reentry vehicles
5. "de-creasing" nuclear arsenals
6. "de-fending" against ballistic missiles
7. "de-weaponizing" outer space
8. "de-militarizing" fissile materials
9. "de-minishing" the inherited nuclear infrastructure

The Trouble With Zero. Slightly paraphrased below is an impressive contemporary summary by Phillip Taubman, published in the *New York Times* (9 May 2009), about problems in getting to zero nuclear weapons. (Missing from this, as from many other related analyses, is recognition of the need to degrade fissile materials.)

Almost from the moment the first atomic bomb was detonated in New Mexico in July 1945, the menacing aura of the nuclear age has inspired visions of a world free of nuclear weapons. Never more so than now, with the prospect that the Taliban could someday control Pakistan's nuclear weapons, North Korea might develop nuclear-tipped missiles, Iran may soon become a nuclear power and terrorists could get a bomb.

A growing army of nuclear abolitionists, concerned that proliferation could catch fire at any moment, is advancing the cause, led by Barack Obama, the first president to make nuclear disarmament a centerpiece of American defense policy....

Yet even as the allure of disarmament grows, the obstacles seem as daunting as ever. Going to zero, as the nuclear cognoscenti put it, is a deceptively simple notion; just about everyone who knows nuclear weapons agrees it would be wickedly difficult to achieve.

That's because it would require a sea change in a dizzying array of defense matters, ranging from core defense policies to highly technical weapons programs. To fully grasp the political and military implications, consider what would have been involved had the great powers of the 19th century decided to abolish gunpowder.

Mr. Obama acknowledges that getting to zero won't be easy. "The goal will not be reached quickly — perhaps not in my lifetime," he declared last month before a huge crowd in Prague. "It will take patience and persistence."

But like other proponents, Mr. Obama has made the eradication of nuclear weapons a pivotal goal, no matter how distant, to provide a lodestar for world leaders and citizens alike.

The new appeal of an old idea that long seemed quixotic is driven by the rise of new nuclear threats that in some ways make the nuclear equation more ominous and volatile than during the Cold War, even though there are far fewer weapons now. Mr. Obama said it himself in Prague: "In a strange turn of history, the threat of global nuclear war has gone down, but the risk of a nuclear attack has gone up."

Nuclear conflict between the United States and the Soviet Union was a prospect so harrowing that American and Soviet leaders recognized it was untenable, even as their generals planned for Armageddon. They possessed some 70,000 nuclear warheads between them in the 1980s, but the weapons were under firm control and neither side dared risk the retaliation that a first strike would draw. The balance of terror, in effect, neutralized nuclear weapons.

The dynamic today is much less stable, and more difficult for the United States to manage, as the turbulence in Pakistan shows. As the nuclear club expands, the security of weapons and technology diminishes. Terrorists would have no compunction about using a nuclear weapon, and their target could not easily retaliate against an elusive, stateless group.

Faced with these dangers, Mr. Obama is banding with fellow leaders like President Dimitri Medvedev of Russia and Gordon Brown, the British prime minister, to push for a series of steps to reduce nuclear threats in the near term, while preparing ground for the eventual elimination of nuclear weapons.

The Obama administration and other advocates favor a reduction in American and Russian nuclear arsenals, to be followed by talks that include nations with smaller nuclear arsenals, like China. They want the United States Senate to ratify the 1996 Comprehensive Test Ban Treaty; would strengthen the 1968 Nuclear Non-Proliferation Treaty; and would seek an accord to verifiably ban the production of fissile materials intended for use in nuclear weapons.

Sam Nunn, the former chairman of the Senate Armed Services Committee, likens such steps to building "a base camp" that offers "a vantage point from which the summit is visible and the final ascent to the mountaintop is achievable." It is an audacious agenda, but as alarm about nuclear threats rises, the chances of success seem to be growing, at least for some interim steps.

Past efforts have foundered. A 1946 plan named after the American financier Bernard Baruch died partly because its scheme to have a powerful international agency control nuclear technology required the five permanent members of the United Nations Security Council to give up their veto power on some nuclear matters. The Nuclear Non-Proliferation Treaty, 41 years old now, has proved ineffectual in moving the world toward nuclear disarmament.

Ronald Reagan and Mikhail Gorbachev briefly considered eliminating nuclear weapons, during their 1986 summit meeting in Reykjavik, Iceland. The idea died when Mr. Reagan refused to abandon his missile defense program.

Mr. Gorbachev, still pushing hard for nuclear disarmament 23 years later, co-hosted an international conference on nuclear issues in Rome last month, a few weeks after Mr. Obama was in Prague. Mr. Gorbachev noted that nuclear disarmament would be untenable to many nations if it left America with overwhelming superiority in conventional military forces. That is one of the biggest potential sticking points.

Nuclear disarmament would also upend decades of American defense strategies. Since early in the cold war, they have been pinned to the chilling concept that a nuclear attack on the United States, and perhaps a chemical or biological attack, would be answered with a devastating nuclear strike.

Dismantling America's nuclear deterrence strikes many defense experts as unwise, if not suicidal. They ardently believe that nuclear weapons have made global war less likely. Harold Brown, a former defense secretary, and John Deutch, a former C.I.A. director — both for Democratic presidents — argue that America will long need a potent nuclear arsenal for deterrence.

They also suggest that the goal of eliminating nuclear weapons is a distraction from, rather than an impetus to, more modest but significant steps to reduce nuclear threats. To counter such concerns, Mr. Obama promised in Prague that as long as nuclear weapons exist, America would "maintain a safe, secure and effective arsenal to deter any adversary" and defend allies.

How far can nuclear arms levels be reduced short of abolition while still providing deterrence? The United States and Russia are opening talks that seem likely to bring the number of operational strategic warheads on each side down to 1,500, possibly 1,000, from the present 2,200.

Those numbers are generally deemed ample for deterrence. But the limit might have to go to 500 or fewer before nuclear weapons states with smaller arsenals, including China, would start cutting. And thousands of American and Russian tactical nuclear weapons, designed for battlefield use, would have to be eliminated, too. At those levels, there is intense debate about whether American security would be gravely undermined.

One solution suggested by abolition advocates would be a form of latent or virtual deterrence, based not on weapons all but ready to launch, but on the ability to reassemble or rebuild them.

If arsenals are drastically reduced, the next steps toward abolition could be even trickier. Since scientific and engineering knowledge cannot be expunged from mankind's memory, the potential to build weapons will always exist. Efforts to hide a few weapons may be difficult to detect and prevent. And any nation able to enrich uranium usable in nuclear power plants, like Iran, has a capacity to produce highly enriched fuel for weapons. Nuclear arms experts have been analyzing these issues intently and have come up with plans to address them. The steps include improvements in the tools used to monitor and verify compliance with treaties and new ways to prevent cheating, including more intrusive inspections.

The enrichment problem, they say, could be solved by limiting the production of enriched uranium to internationally controlled fuel banks that would supply power reactors in places like Iran, eliminating the need for national enrichment plants.

The notion of nuclear disarmament gained credibility a few years ago when four cold war veterans — George Shultz and Henry Kissinger, former secretaries of state; William Perry, a former defense secretary; and Mr. Nunn — overcame their political differences to endorse the idea in a Wall Street Journal op-ed article. Now that it has been embraced by Presidents Obama and Medvedev, the notion seems to be moving from the realm of fantasy to the hardscrabble world of policy and politics.

****

Some frightening aspects of history can repeat themselves, especially if incongruous or irrelevant lessons are absorbed by new generations of policymakers. Fear of yet another superpower nuclear confrontation cannot be simply dismissed — at least not as long as the wherewithal lingers.

Although a logical case can be made for nuclear weapons having sustained a 45-year *pace regnum*, there is little prospect in the 21st century that nuclear arsenals would be helping the cause of peace, nor improving universal living conditions, nor even quelling terrorism.

The lead author of this book and his colleagues are part of what might be called the "coexisting generation:" During our time, the superpowers coexisted in a tense and protracted nuclear stalemate. It falls upon us to pass on a vivid institutional memory, hoping to mitigate the mass-casualty hazard for future generations. The nuclear option can be diminished: partly by recalling the dangers and lessons of the Cold War, as this book does, and mostly reducing temptations and implements of war. Phasing out as many nuclear weapons, weapon-grade materials, and nuclear-weapon labs as possible would accelerate relief from future hazards.

Military strategist General Lee Butler observed that the superpowers consistently misread "each other's intentions, motivations, and activities." Even more ominous is his warning that "Their successors still do so."

[www.GlobalZero.org]. Paris (December 2008).

In response to the growing threats of proliferation and nuclear terrorism, 100 leaders from around the world launched Global Zero. They for the phased, verified elimination of nuclear weapons, starting with deep reductions in the U.S. and Russian arsenals, to be followed by multilateral negotiations among all nuclear powers for an agreement to eliminate all nuclear weapons — global zero. The growing group includes former heads of state, former foreign ministers, former defense ministers, former national security advisors, and more than 20 former top military commanders. We committed our two countries to achieving a nuclear-free world, while recognizing that this long-term goal will require a new emphasis on arms-control and conflict-resolution measures, and their full implementation by all concerned nations.

Global Zero is working on three fronts to achieve this goal: 1) developing a step-by-step plan for elimination of nuclear weapons based on the Paris conference framework; 2) conducting track-two diplomacy to build support among key governments; and 3) generating broad-based worldwide public support through media and online communications and grassroots organizing.

Global Zero will convene hundreds of international leaders for the Global Zero Summit in February, 2010.

One of our biggest concerns, a driving force for this book, has been that so few lessons have been tangibly learned, appreciated, or remembered. We are apprehensive that coming generations of leaders, especially "baby boomers," will not recall the dangers of such brinkmanship. Nor might they appreciate the environmental and cultural legacy that remains to be addressed or the necessity of scrupulously controlling and reducing nuclear arms and fissile materials.

Cold War strategic-policy was sometimes parsed in terms of abstract algorithms: *compellence, deterrence, or avoidance*. Compellence was (and still is) pushed by hardliners; deterrence was the prevailing compromise; avoidance was fostered by grass-roots public activism.

But now, it's time for *nuclear de-emphasis*, as summarized in our preceding list of key phase-out measures, justified through this book's three volumes.

During our professional careers, my colleagues and I gained hands-on experience in nuclear development and treaty verification, mainly with short-term goals. Now retired from professional employment, this is our thoroughly considered contribution toward longer-term de-emphasis aimed at eventual renunciation of nuclear warfare.

Gradual nuclear reductions, accompanied by confidence-building verification, are needed as an interim bridge to a less-challenged future. With fewer nuclear weapons and materials, concomitant security problems would gain immense relief. If the non-proliferation norm were reinforced, it would have a chance of becoming universal. Moreover, the benefits derived from modern technology, especially nuclear, could be put to better and more constructive use for society.

While nations retain adequate conventional-defense capabilities, verified arms-control measures are essential and feasible to prevent rapid revitalization of nuclear weapons, materials, infrastructure, and laboratories.

Alleviation of conflict and deep reductions will not occur unless economic and social institutions of nations and regions gain structural stability and security, along with adequate resources.

Moreover, as we learned with the atomic bomb, new technologies could create potentially severe problems for future societal restraint and adaptation.

****

Realizing that human memories fade, it is evermore convenient to have assembled in one compilation a comprehensive assessment of the nuclear arms race. If for no one else's benefit but our own, we have digested and documented personal and vicarious experiences.

Though my colleagues and I were bit players in the coexisting generation, we used inside experience and published resources to fill in historical and analytical gaps. As physicists, we have tried to be systematic in our undertaking, while sidelining our personal whims and biases. Through diversified authorship and outreach, we have endeavored to bridge huge cultural gaps. By applying traditions of scientific discipline to this composition, we hope to have produced a valuable and enduring contribution.

None of us wants our respective native or adoptive nation to let down its guard, so a step-by-step transition to a safer arms balance is essential for thwarting disaster.

During the writing of this book, as issues popped up, a sense of *déjà vu* has frequently reoccurred: Wasn't the mutual-defense initiative discussed back in the days of the SDI or the BDM? Or the ABM? Hasn't nuclear temptation been a topic of concern for many decades?

Weren't mini-nukes problematic during the NATO-nuclearization debates? Isn't the current controversy over nuclear, chemical, or biological terrorism a rehash of the past?

Current-day proliferation concerns sound familiar. Weren't ways of ridding ourselves of weapons-grade plutonium suggested decades ago? Why are we still rehashing nuclear-waster disposal issues? *Deja vu* indeed!

Though not an end-all in itself — there is too much to say and too many other voices that must be heard — we shaped our original collaborative book, *Nuclear Shadowboxing*, to be an enduring reference for this menacing phase of contemporary history. *Nuclear Insights*, is a briefer, more readable interpretation of the complex historical and technical events.

This third Volume offers a technically informed perspective about proposed arms reductions. That perspective is derived upon the our knowledgeable assessment of inherited nuclear threats and prospects presented in Volume 2, which is founded on our insider history of Cold War nuclear weaponry, narrated in Volume 1.

Importantly, we have tried to adhere to our own standards and practice of scientific reserve: that is, the recognition of statistical uncertainty for any type of analysis, including that which becomes the basis for policy decisions. While we have made recommendations, they are subject to appropriate bounds of context and incertitude.

In searching for familiar names of individuals now in the federal government and in influential organizations, the absence of technical people is noticeable. Yet, while politicians and economists and political scientists run the government, there are some matters that require input from technologists, especially those who have had hands-on experience and a history of involvement with nuclear issues, particularly an institutional memory of the Cold War: That's us, when it comes to shepherding the way for national and multilateral security after nuclear realignment.

In any event, everybody should remember, nuclear weapons never won a war, but their holders and beholders continue to jeopardize peace and accommodation in the new millennium. That realization is perhaps our capstone.

When you finish reading this book, hopefully humanity will be awakening from its nightmare about nuclear weapons and nuclear warfare.

###

[UN Secretary-General Ban Ki-moon's "Plan to Stop the Bomb"].

The world is at a turning point – nuclear disarmament is back on the global agenda.

The Cold War's end, 20 years ago this autumn 2009, was supposed to provide a peace dividend. Instead we find ourselves still facing serious nuclear threats. Some stem from the persistence of more than 20,000 nuclear weapons and the contagious doctrine of nuclear deterrence. Others relate to nuclear tests – more than a dozen in the post-cold war era, aggravated by the constant testing of long-range missiles. Still others arise from concerns that more countries or even terrorists might be seeking the bomb.

For decades, we believed that the terrible effects of nuclear weapons would be sufficient to prevent their use. The superpowers were likened to a pair of scorpions in a bottle, each knowing a first strike would be suicidal. Today's expanding nest of scorpions, however, means that no one is safe.

Many efforts are under way worldwide to achieve this goal. Earlier this year, the Conference on Disarmament broke a deadlock and agreed to negotiations on a fissile material treaty. Other issues it will discuss include nuclear disarmament and security assurances for non-nuclear-weapon states.

Next May (2010), the UN will host a major five-year review conference involving the parties to the Nuclear Non-Proliferation Treaty (NPT), which will examine the state of the treaty's "grand bargain" of disarmament, non-proliferation and the peaceful use of nuclear energy. If the CTBT can enter into force, and if the NPT review conference makes progress, the world would be off to a good start on its journey to a world free of nuclear weapons.

My own five-point "musts" for disarmament:
- reliable verification
- enhanced security
- rooted in legal obligations
- visible to the public
- anticipate emerging dangers from other weapons

Figure 9: A Night-Time View of Post-Millennium Hiroshima

ACRONYMS AND ABBREVIATIONS

AAAS	American Association for the Advancement of Science
AAM	air-to-air missile
ABACC	Brazilian-Argentine Agency for Accounting and Control of Nuclear Materials
ABM	anti-ballistic missile
ACA	Arms Control Association
ACDA	Arms Control and Disarmament Agency (U.S.)
ACHRE	Advisory Committee on Human Radiation Experiments
ADM	atomic demolition munition
AEDS	Atomic Energy Detection System (network for detecting nuclear explosions)
AFB	Air Force Base
AFTAC	Air Force Technical Applications Center (performs treaty-verification functions)
ALCM	air-launched cruise missile
ANL	Argonne National Laboratory
ANP	aircraft nuclear propulsion
ANWFZ	African Nuclear Weapons Free Zone Treaty (Treaty of Pelindaba)
APS	American Physical Society
ASAT	antisatellite
ASM	air-to-ship missile
ASROC	antisubmarine rocket (launched from a surface ship). See SUBROC
ASW	anti-submarine warfare
BAMBI	Ballistic Missile Boost Intercept
BMD	ballistic missile defense
BMDO	Ballistic Missile Defense Organization (predecessor to the MDA)
BMEWS	Ballistic-Missile Early-warning System (U.S. radars in Alaska, Greenland, and England)
BNCT	Boron neutron capture therapy
BWC	Biological Weapons Convention (entered into force 1975)
C&C	command and control
C^3	command, control, and communications
C^3I	command, control, communications, and intelligence
CAS	Concerned Argonne Scientists
CBO	Congressional Budget Office
CDI	Center for Defense Information
CEP	circle of error, probable — the radius of the circle, centered on the target, within which a warhead has a 50% chance of landing
CFE	Conventional Forces in Europe (treaty signed in 1990)
CIA	Central Intelligence Agency (U.S.)
CIS	Commonwealth of Independent States (former republics of the Soviet Union)
CISAC	Center for International Security and Cooperation (at Stanford University) (formerly Center for International Security and Arms Control)
CISSM	Center for International Security Studies (at the University of Maryland)
CND	Campaign for Nuclear Disarmament (England)
COCOM	Coordinating Committee on Multilateral Ex-port Controls
COMECON	Council for Mutual Economic Assistance — an economic organization of communist countries
CPD	Committee on the Present Danger

CPSU	Communist Party of the Soviet Union
CSBM	confidence-and security-building measures
CSCE	Conference on Security and Cooperation in Europe
CSS	Committee of Soviet Scientists for Peace and Against the Nuclear Threat
CTR	Cooperative Threat Reduction
CWC	Chemical Weapons Convention (entered into force 1997)
CTBT	Comprehensive Test Ban Treaty (rejected by the U.S. Senate, 13 October 1999)
D-5	Trident II D-5 (a U.S. MIRVed SLBM)
DARPA	Defense Advanced Research Projects Agency (U.S.)
DIA	Defense Intelligence Agency (U.S.)
DNA	Defense Nuclear Agency (now DSWA) (U.S.)
DOD	Department of Defense (U.S.)
DOE	Department of Energy (U.S.)
DSWA	Defense Special Weapons Agency (formerly DNA) (U.S.)
ELF	extremely low frequency
EMP	electromagnetic pulse (caused by a nuclear explosion)
END	European Nuclear Disarmament (NGO peace movement)
ENDS	enhanced nuclear detonation system (reduces the chance of a warhead's detonators being fired electrically in an accident)
EPNW	earth-penetrating nuclear weapon
ERIS	Exoatmospheric Reentry Interceptor Subsystem (ballistic missile interceptor)
ERW	enhanced radiation warhead (the "neutron bomb" is an ERW)
FAS	Federation of American Scientists
FDA	Food and Drug Administration (U.S.)
FEMA	Federal Emergency Management Agency (U.S.)
FMCT	Fissile Material Cut-off Treaty (also called "fissban")
FOBS	fractional-orbit ballistic systems
FOIA	Freedom of Information Act (U.S.)
FRG	Federal Republic of Germany (West Germany)
FRP	fire-resistant pit (a nuclear-warhead safety feature for withstanding prolonged exposure to a jet-fuel fire)
FSU	former Soviet Union
FY	fiscal year
GAO	General Accounting Office — the audit, evaluation, and investigative arm of the U.S. Congress
GEO	Geosynchronous orbit (satellite)
GLCM	ground-launched cruise missile
GPS	Global Positioning System (U.S.) (Navstar), consisting of dozens of satellites
GRU	The Soviet Main Intelligence Administration, operated by the Army
HE	conventional high explosive, used in early designs of U.S. nuclear weapons
HEO	Highly elliptical earth orbit (satellite)
HEU	highly enriched uranium
HOE	Homing Overlay Experiment — a missile-interception test series
HTDS	Hanford Thyroid Disease Study
HUAC	House Un-American Activities Committee (U.S.,1934-1977)
HUMINT	human means of information gathering
IAEA	International Atomic Energy Agency
ICBM	intercontinental ballistic missile
ICF	inertial-confinement fusion
IGCC	Institute on Global Conflict and Cooperation (at University of California San Diego)
IHE	insensitive high explosive (a nuclear-warhead safety feature) used in some of the more recently designed warheads

IMEMO	Institute for World Economy and International Relations (Moscow)
IMS	International Monitoring System (for the CTBT, if ratified)
INESAP	International Network of Engineers and Scientists Against Proliferation
INF	intermediate-range nuclear forces
INSTEAD	Interuniversity Network for Studies on Technology Assessment in Defence (at the Freije University of Amsterdam)
IPPNW	International Physicians for the Prevention of Nuclear War
IRBM	intermediate-range ballistic missile
ISODARCO	International School on Disarmament and Research on Conflicts
JCAE	Joint Committee on Atomic Energy (U.S. Congress, 1947–1977)
JCS	Joint Chiefs of Staff (U.S.)
JDEC	Joint Data Exchange Center, in Moscow, "to ensure the uninterrupted exchange of information on the launches of ballistic missiles and space launch vehicles." Established June 2000; suspended by the G.W. Bush administration
KGB	Committee for State Security (Soviet) — tasked with ferreting out potential threats to the state and preventing the development of unorthodox political and social attitudes among the population [*FAS Web site*]
LANL	Los Alamos National Laboratory
LEO	Low earth orbit (satellite)
LEU	Low-enriched uranium
LNT	Linear No-Threshold—the theory that biological damage by radiation extends linearly all the way down to zero dose.
LoAD	low-altitude defense
MAD	mutual assured destruction
MBFR	Mutual Balanced Force Reduction (in Europe)
MDA	Missile Defense Agency (formerly the BMDO)
MED	Manhattan Engineering District (the Manhattan Project)
MEO	mid-earth orbit (satellite)
MeV	million electron volts (energy unit for sub-atomic particles)
MILC	military-industrial-laboratory complex
MINATOM	The Ministry for Atomic Energy of the Russian Federation
MIRACL	Mid-Infrared Advanced Chemical Laser (a deuterium-fluoride chemical laser)
MIRV	multiple, independently targetable re-entry vehicle
MIT	Massachusetts Institute of Technology
MOD	Ministry of Defense (Russia)
MOX	mixed-oxide fuel for power reactors (blend of oxides of uranium and plutonium)
MPC&A	Materials Protection, Control, and Accountability
MRBM	medium-range ballistic missile
MRV	multiple re-entry vehicle (not independently targetable)
MTCR	Missile Technology Control Regime (an agreement among missile suppliers)
MWe	Megawatt-electric
MX	Missile-Experimental (a U.S. ten-warhead ICBM, dubbed "Peacekeeper" by Ronald Reagan)
NAOC	National Airborne Operations Center (U.S.)
NASA	National Aeronautics and Space Administration (U.S.)
NATO	North Atlantic Treaty Organization
NCRP	National Council on Radiation Protection and Measurement
NGO	non-governmental organization
NIE	National Intelligence Estimate
NIS	Newly Independent States (of the former Soviet Union)
NMD	national missile defense
NNWS	non-nuclear-weapons state(s)
NORAD	North American Aerospace Defense Command

NOMOR	**N**uclear **O**verkill **MOR**atorium — a Chicago-based anti-arms-race organization
NPR	Nuclear Posture Review (quadrennial U.S. military review)
NRC	Nuclear Regulatory Commission (U.S.)
NRDC	Natural Resources Defense Council
NPR	Nuclear Posture Review (initiated by the Clinton administration, 1994)
NPT	Non-Proliferation Treaty (1968)
NSA	National Security Agency (U.S.)
NSC	National Security Council (U.S.) — a Presidential advisory body
NSDM	National Security Decision Memorandum
NTM	national technical means (for detecting the activities of an adversary)
NTS	Nevada Test Site
NWFZ	nuclear-weapon-free zone
NWS	nuclear-weapons state(s)
OPS	one-point safe (a nuclear warhead safety-design feature)
OSI	on-site inspection
OSS	Office of Strategic Services (U.S.)
PAL	permissive action link (locking device to prevent unauthorized firing of a nuclear weapon)
PD	Presidential Directive
PNE	peaceful nuclear explosion; Peaceful Nuclear Explosions treaty (signed 1976; ratified 1990)
PSAC	Presidential Scientific Advisory Committee
PTBT	Partial Test Ban Treaty
PSR	Physicians for Social Responsibility
R&D	research and development
RBE	relative biological effectiveness (of ionizing radiation)
RERTR	Reduced Enrichment for Research and Test Reactors program (at Argonne National Laboratory)
RF	Russian Federation
RV	re-entry vehicle (the unit containing the warhead as it re-enters the atmosphere)
RV	remotely piloted vehicle
SAC	Strategic Air Command (now SC) (U.S.)
SALT	Strategic Arms Limitation Treaty, for which negotiations between the U.S. and the SU started in November 1969, leading to the treaties SALT-I (1972) and SALT-II (1979).
SAM	surface-to-air missile
SANE	Committee for a SANE Nuclear Policy
SC	Strategic Command (formerly SAC) (U.S.)
SCC	Standing Consultative Commission (set up to resolve Salt-I disputes)
SDI	Strategic Defense Initiative (the "Star Wars" BMD program started by President Reagan in 1983)
SESPA	Scientists and Engineers for Social and Political Action (also known as "Science for the People")
SIGINT	signals intelligence
SIOP	Single Integrated Operational Plan (U.S. list of potential nuclear targets)
SIOP-ESI	SIOP-Extremely Sensitive Information (a U.S. secrecy classification for SIOP information)
SIPRI	Stockholm International Peace Research Institute
SLBM	submarine-launched ballistic missile
SLCM	submarine-launched cruise missile
SNDV	strategic nuclear delivery vehicle
SNTRA	Soviet Nuclear Threat Reduction Act (U.S.), passed in 1991

SPOT	Satellite Pour l'Observation de la Terre (a French high-resolution imaging satellite system)
SRAM	short-range attack missile
SRBM	short-range ballistic missile
SSBN	ballistic-missile submarine, nuclear-powered
SSM	surface-to-ship missile *or* ship-to-ship missile
START	Strategic Arms Reductions Talks (successor to SALT)
SUBROC	submarine rocket (anti-submarine weapon launched underwater from a submarine's torpedo tube). See ASROC
	transporter-equipment-launcher (for land-mobile missiles)
THAAD	Theater High-Altitude Aerial Defense (also called Theater High-Altitude Area Defense)
TTBT	Threshold Test Ban Treaty (signed 1974; ratified 1990)
TLI	treaty-limited item
UCS	Union of Concerned Scientists
UK	United Kingdom
UN	United Nations
UNIDIR	United Nations Institute for Disarmament Research
UNMOVIC	United Nations Monitoring, Verification and Inspection Commission (established Jan.1990 to inspect Iraq for WMD)
UNSCOM	United Nations Special Commission, set up in 1991 to help the IAEA deal with Iraq's weapons programs
USEC	United States Enrichment Corporation
USSR	Union of Soviet Socialist Republics (Soviet Union)
VERTIC	Verification Research, Training & Information Centre (formerly the Verification Technology Information Centre)
WMD	weapons of mass destruction
WTO	Warsaw Treaty Organization

LIST OF FIGURES

GLOSSARY

atomic bomb	A bomb that uses a fission chain reaction to create a devastating explosion that results in blast, heat, and radiation effects
atomic weight	The number of protons plus neutrons in the atomic nucleus
beltway bandits	Contractors in the Washington, DC, area who seek government contracts
boosting	A process of increasing the explosive yield of a nuclear weapon by inducing fusion reactions
brinksmanship	The pursuit of an imperious and dangerous policy to the boundary between safety and recklessness
broken arrows	Accidents involving nuclear weapons that do not cause a nuclear explosion
calutron	An early means of enriching uranium by bending a beam of uranium ions in a magnetic field
chain reaction	A series of nuclear fissions linked by the neutrons released
counterforce	A policy of attacking military rather than civilian targets — opposite of *countervalue*
countervalue	Application of military force against targets of economic value, rather than military targets
critical mass	The minimum amount of nuclear material that supports a self-sustaining chain reaction
criticality	A condition reached when a mass of fissile material sustains a nuclear reaction
curie	An amount of radioactive material that decays at the rate of 3×10^{10} disintegrations per second
deterrence	A policy of dissuasion by threat of military retaliation
doomsday machine	An apparatus that endangers all of civilization
fast neutrons	Neutrons that have comparatively high velocity, such as those released from a fissioning nucleus

fertile isotope	An isotope whose atoms become fissile after absorbing one neutron; e.g., Th-232, U-238, Pu-240
first strike	A (nuclear) attack made without warning
first-strike capability	The ability to launch such a devastating attack that the opponent's retaliatory means is destroyed; the opponent therefore is essentially denied a *second-strike capability*.
fissile isotope	An isotope that can be fissioned by thermal neutrons; e.g., U-233, U-235, Pu-239, Pu-241
fissionable isotope	An isotope (either fissile or fertile) that can be fissioned by fast neutrons; all fissile atoms are fissionable, but not all fissionable atoms are fissile
fission (nuclear)	An energetic breakup of a nucleus, resulting in the release of neutrons, other radiation, and energy
fission products	Lighter elements formed when nuclear fission occurs — the "ashes" of the fission process
fission weapon	A weapon based on an uncontrolled nuclear chain reaction
fusion (nuclear)	Nuclear reactions in which nuclei are forced (fused) rapidly together to form heavier elements, plus release of energy and radiation
fusion weapon	A weapon that derives much of its energy from nuclear-fusion reactions
gun-barrel weapon	A nuclear weapon design in which two subcritical fissile masses become supercritical when slammed together (rapidly compressed) inside a tube
glasnost	Openness (Russian)
Gulf War	The war in the Persian Gulf that began with Iraq's invasion of Kuwait in 1990 and ended after a UN-coordinated invasion of Iraq in 1991 (Operations Desert Storm and Desert Shield)
horizontal proliferation	An increase in the number of nations that develop or obtain nuclear weapons
implosion	A process in which fissile materials are rapidly compressed inward in order to quickly reach explosive criticality

initiator	A means of generating neutrons within a nuclear assembly in order to begin the chain reaction
Iraq Invasion	The invasion of Iraq in 2003 by a U.S.-led coalition (operation Iraqi Freedom)
isotope	A nuclear species consisting of a specific number of protons (the atomic number of the element), and a specific atomic weight; e.g., two of the isotopes of uranium (atomic number 92) are U-235 and U-238
natural uranium	Uranium as found in nature, usually having 99.3% fissionable U-238 and 0.7% fissile U-235
neutron weapon	A nuclear weapon that sacrifices explosive yield in order to emit more neutrons and other radiation
nuclear	Pertaining to the atomic nucleus
nuclear weapon	A fission or fusion weapon that is designed and qualified for use by military forces
pit	The central part of a nuclear weapon — the fissile core
perestroika	Restructuring (Russian)
production reactor	A nuclear reactor operated so as to produce weapons-grade plutonium
proliferation	An increase in the number of nations that have nuclear weapons — horizontal proliferation
radiation	Energy emission from an atom or its nucleus in the form of rays or particles
reactor-grade plutonium	Plutonium discharged from a "thermal-spectrum" nuclear reactor. It could have a fissile-isotope concentration of up to 80%, well below the requirements for military-quality nuclear explosives (see weapons-grade plutonium).
reprocessing	Treating spent fuel from a nuclear reactor so as to be able to reuse some of the fissionable components after separating them from fission products and other materials
second-strike capability	The ability to make an assured and devastating response after absorbing a first-strike attack

secondary	The second stage of a thermonuclear weapon in which most of the fusion reactions take place
strategic (nuclear) weapons	Nuclear weapons deliverable over long distances
tactical (nuclear) weapons	Nuclear weapons intended for battlefield use — variously called theater, battlefield, non-strategic, or sub-strategic weapons
thermal neutrons	Neutrons that have slowed down until they are in approximate thermal equilibrium with their surroundings
thermonuclear weapon	Another name for a fusion weapon, usually an explosive device that consists of a fission trigger and a separate fusion stage to amplify the explosive yield
throw-weight	The total load (usually RVs with nuclear warheads) that can be delivered to a target by a ballistic missile
vertical proliferation	An increase in the number of nuclear weapons in a nation's arsenal
weapons-grade	Nuclear material that meets specifications for manufacture of a military-quality nuclear weapon
weapons-grade plutonium	Plutonium that contains at least 94% fissile isotopes (primarily Pu-239) — the balance being mostly fertile isotopes (primarily Pu-240)
weapons-grade uranium	Uranium that contains at least 93% fissile isotopes (usually U-235) — the balance being the fertile isotope U-238

AUTHOR BIOGRAPHY

Four nuclear scientists participated in the original publication (Nuclear Shadowboxing) *that gave rise to the three volumes of* Nuclear Insights: *Dr. Alexander DeVolpi, arms-control physicist and lead author; Dr. Vladimir E. Minkov, a physicist and émigré from the Soviet Union who has had a foot on both sides of the cultural divide; Russian coauthor, Dr. Vadim A. Simonenko, weapons physicist who provides am insider's Soviet perspective; and Dr. George S. Stanford, also a physicist, coauthor, and editing specialist. Credentials for all four are supplied in Volume 1.*

DeVolpi alone is responsible for condensing the material into Nuclear Insights.

Dr. Alexander DeVolpi

Alexander DeVolpi has been an arms-control physicist, active in nuclear-arms policy and treaty-verification technology studies, for over 25 years. Now retired from Argonne National Laboratory, he writes from first-hand experience on most topics of this book.

On the subject of nuclear-weapons nonproliferation, DeVolpi is author or coauthor of many articles, two books (author of *Proliferation, Plutonium, and Policy* and co-author of *Born Secret: The H-bomb, The Progressive Case, and National Security* — both published by Pergamon), and a number of technical review articles (including "Fissile Materials and Nuclear Weapons Proliferation," *Annual Review of Nuclear and Particle Science*).

Dr. DeVolpi has initiated numerous projects on the methodology and technology of treaty verification, including a technique for relatively unintrusive counting of nuclear-warhead multiplicity, described in a 1970 paper. Other proposals have pertained to nuclear-warhead detection and inspection on Earth and in space, fissile-material conversion, nuclear-facility monitoring, aerosol applications, weapons dismantlement, tagging and sealing, chemical-weapons verification, laser-brightness monitoring, cargo and luggage inspection, contraband-drug detection, and cooperative treaty-verification measures. He has given papers on these subjects at national and international conferences in Paris (1986), Japan (1989), Germany (1989), Austria (1990), Italy (1991), Canada (1991), Russia/Ukraine (1991), and Geneva (1993).

After earning an undergraduate degree in journalism from Washington and Lee University, DeVolpi served on active duty with the U.S. Navy in the mid-1950s. Later he received his Ph.D. in physics (and MS in nuclear engineering physics) from Virginia Polytechnic Institute, Blacksburg, Virginia.

DeVolpi has considerable experience in neutron physics and nuclear diagnostics, having served as principle investigator in a variety of research projects. He was manager of nuclear diagnostics in the Reactor Analysis and Safety Division at

Argonne, and then became technical manager of the arms-control and nonproliferation program.

He holds a half-dozen patents, one of which is for the neutron/gamma hodoscope, a major instrument system used in the United States and France to image the motion of fissile material which was being tested under simulated accident conditions in special transient reactors.

In Argonne's Arms Control and Nonproliferation Program, he contributed to various technical arms-control projects, becoming Technical Manager for Physics and Engineering, and Principal Investigator for tamper-resistant tags and seals and assessment of foreign verification technology.

DeVolpi is recognized in *Who's Who in Frontiers of Science and Technology, American Men and Women of Science, Who's Who in Science and Engineering*, and other biographies. He was elected a Fellow of the American Physical Society for his contributions to arms-control verification and public enlightenment on the consequences of modern technology. He has been a member of the American Association for the Advancement of Science, the American Nuclear Society, and the Institute for Nuclear Materials.

Gaining the rank of Lieutenant Commander (retired) in the U.S. Naval Reserve as a result of 17 years on active duty and in the reserves, DeVolpi has had numerous assignments to the Naval Research Laboratory and the Radiological Defense Laboratory, where he participated in analysis of fissile-material safeguards and arms-control verification technology — including studies in 1969 regarding detection of ballistic-missile RV multiple-warheads, and monitoring for nuclear weapons on the seabed.

In a technical capacity, he has visited many national and international laboratories and has represented Argonne on various working and advisory groups — in the subjects of arms control, verification technology, radiation detection, tagging, on-site inspection procedures, and ground-based laser verification. He has been funded by the Departments of Energy and Defense to provide and participate in key briefings on related issues in Washington at government and contractor offices.

DeVolpi has also participated in major conferences on verification at Livermore, Sandia, Los Alamos, and Argonne; the DOD/DNA Arms Control Roundtables; American Physical Society, Association for the Advancement of Science, SAIC/CSNS verification workshops; DNA International Conference on Arms Control Verification Technology; and many international symposia. Because of long-term participation in treaty verification, he has had interactions with a wide spectrum of resource persons in and out of government, especially with arms-control organizations.

As a citizen-scientist, Alex has been active in public-interest arms-control issues for over 35 years. He was a participant and technical consultant in the FAS/NRDC joint project with the Soviets on nuclear-warhead dismantlement. He served as an elected member of the national council of the Federation of American Scientists in 1988-92. He was co-founder of Concerned Argonne Scientists, and a member of activist organizations and executive committees in the Chicago area. Some of his public-interest writings drew adverse reactions from DOE classification officials, resulting in a couple of publicized suspensions of his security clearance and the confiscation of his work computer. All of those actions were reversed after higher-level review.

From a previous marriage, Alex had four children, now grown up, two of whom have their own families. He keeps in shape by playing handball, ping-pong, and swimming; and he enjoys — time permitting — poker, fishing, and woodworking as hobbies. Because of the time spent on *Nuclear Insights*, two book projects about his family history and genealogy are backlogged. He now lives in California, north of San Diego.

CONSOLIDATED
TABLE OF CONTENTS

TABLE OF CONTENTS FOR VOLUME 1

TABLE OF CONTENTS FOR VOLUME 2

TABLE OF CONTENTS FOR VOLUME 3

INDEX FOR VOLUME 3

(Hyphenated terms are listed as though the hyphen followed the letter z;
e.g., Arms-Control Agreements *follows* Arms Reduction)

www.ingramcontent.com/pod-product-compliance
Lightning Source LLC
Chambersburg PA
CBHW050110280326
41933CB00010B/1040